James McLean, William Henry Harrison Hart

Treatise on the Origin of Destructive Insect Plagues

and improvements in the art of their eradication and prevention

James McLean, William Henry Harrison Hart

Treatise on the Origin of Destructive Insect Plagues
and improvements in the art of their eradication and prevention

ISBN/EAN: 9783337140373

Printed in Europe, USA, Canada, Australia, Japan

Cover: Foto ©berggeist007 / pixelio.de

More available books at **www.hansebooks.com**

"The leaves of the tree WERE for the healing of the nations."
REVELATION 22. 2.

An Appeal to Every Nation

Gloria in excelsis Deo et in terra pax hominibus bonæ voluntatis.

TREATISE

On the Origin of Destructive Insect Plagues and Improvements in the Art of their Eradication and Prevention. From a Metereological and Hygienic Basis. Revised and re-written in San Francisco, California, U. S. A., in April, 1893, under the auspices of

THE HON. W. H. H. HART,

Attorney-General of the State of California, and author of the famous "Contagious Fruit Trees Ac'," et

—— BY ——

Jas McLean, M. D.

MEDICAL ORCHARDIST

(For twenty years Senior Inspector of Forests and Agricultural Settlement in Victoria, Australia.)

Respectfully Dedicated to the Founder of "Arbor Day,"
THE HON. J. STERLING MORTON,
United States Minister of Agriculture.

WORDS OF WISDOM:

"The strong man who, in the confidence of sturdy health, courts the sternest activities of life and rejoices in the hardihood of constant labor, may still have lurking near his vitals unheeded disease that dooms him to sudden collapse."—PRESIDENT CLEVELAND, (March 4th, 1893.)

{ Letters Patent }
{ applied for }

AMERICAN ADDRESS,
CARE LANGLEY & MICHAELS COMPANY
34 TO 40 FIRST STREET, SAN FRANCISCO, CAL.

{ Price 25 cents }
{ All rights reserved }

Printed for the Publisher by McCormick Bros., 508 Sansome Street, San Francisco, Cal., June, 1893.

PREFACE.

As " there is no new thing under the sun," and being consciou
of my inability to do otherwise than feebly verify that fact, I ha·
ventured to adopt the following philosophic lines penned by LESSING
a preface to this somewhat necessarily fragmentary treatise :   " Eve
man has his own style, as he has his own nose, and it is neither poli
nor Christian to rally an honest man about his nose however singul
it may be.   How can I help it that my style is not different?   Th
there is no affectation in it I am certain."

Jas. McLean, M. O.

San Francisco, June, 1893.

THE
ORIGIN OF DESTRUCTIVE INSECT PLAGUES
—AND—
GENERAL ATMOSPHERIC TROUBLES.
HOW DISCOVERED.
REMEDY.
ETC.

THE EVOLUTION OF OBSCURE TRUTHS.

The African savage when he takes off his fur kaross is familiar with the electric sparks which come from it; but he views them with the eye of an ox, and thinks nothing more about them. The American Indian, in the dry climate of the United States, must constantly have seen these sparks, but never dreamed of making Franklin's experiment by bringing them down from a thunder-storm and showing that they were identical with lightning. The science of electricity and all scientific conceptions arise only when culture develops the human mind and compels it to give a rational account of the world in which man " lives and moves and has his being." One hundred and twenty years before the Christian era, Hero, a renowned mechanician of Alexandria, Egypt, discovered the power of steam when confined in a closed vessel, and he invented the " œlopile," a machine whose arms were propelled by the reaction of issuing jets of steam. It was only an ingenious toy, but it contained " the promise and potency" of the remarkable motor which twenty centuries later re-invented by Papin and Savery, finally received its finishing touches from the fertile brain and cunning hand of my illustrious countryman and townsman, James Watt, who left the steam engine the practically perfected machine of to-day, for whenever improvements have been made they have been on lines laid down by Watt. In like manner the illimitable store house of nature—the fountain of all benificent ideas— has been from time to time explored by the agency of simple " ingenious toys," etc. Sir Isaac Newton, for example, when playing as a little boy on the bank of a favorite stream was led to perceive the hitherto unnoticed, though immutable, law of attraction and repulsion by the agency of sundry little globules of water gambolling in detached particles as they were propelled from the adjacent bubbling brook; how that globules of equal proportions were repellant to each other, whilst the smaller or more negative were attracted to and absorbed by the larger or more positive. It should be needless in these enlightened times by any special reasonings to affirm that this now universally recognized law intimately applies to the whole of organic nature. Man, for instance, who is but a migratory tree or shrub becomes, from various preventable causes, mentally and physically enfeebled, and is in consequence, as are all manner of plants susceptible to attacks from

MINUTE FUNGACIOUS ORGANISMS.

Professor Carl von Naegeli, in his work on " The Minutest Fungacious Organisms in relation to Infectious Diseases and Sustenance of Health," declares that " the decompositions effected by the mould fungi may be decay or consumption. Under their influence, for instance, fruits putrify, or wood is converted into mould by a kind of slow combustion of organic substances. The decompositions to which the sprouting fungi or *saccharomyces* gives rise, are these of fermentation. By their agency sugar is converted into alcohol and carbonic acid. The infectious principles of septic diseases (that which promotes the putrifaction of organic bodies) manifest themselves through putridity, originated again by peculiar fungi, which may be the bearers of a separate putrid matter also. The atmosphere is the medium through the agency of which infectious germs are most generally disseminated after they have been reduced to a state of minute dust by desecation." Other eminent scientists and numerous observing laymen are hourly verifying the above affirmation as they become more familiar with

THE CHEMISTRY OF CREATION.

Professor Ellis, F.L.S., in his standard work on the " Chemistry of Creation," (folio 162) states : " Insects, fish, lichens, infusorial animalcules, volcanic ashes, sand, earth, and many other substances are occasionally borne into the air by the action of rapidly revolving currents, and are dropped often at a great distance from the places whence they were snatched." And concerning our long pitiable ignorance regarding the composition and functions of the air we breathe, Professor Ellis declares (folio 130) " until the middle of the eighteenth century the opinion was very prevalent that the atmosphere formed one of the four elementary bodies; that it was in fact a simple, undecomposable gas. It was reserved for the talented Dr. Priestly to dispel this error. He discovered the existence of a new gas which formed one of the constituents of air. In this gas it was found that combustion took place with extraordinary intensity; even iron wire heated red-hot and plunged into it caught fire and burnt away. Other combustibles gave out showers of the most brilliant sparks and produced the most intense heat when placed in the jar containing it. A lighted taper having been blown out immediately rekindled when put into it and blazed with much greater brilliancy than in the air. Another gas was also found to form a component of the air— namely, nitrogen. The former being oxygen. The writer proceeds to state that " animals were exhilarated when plunged into oxygen, and they were suffocated in nitrogen. "A never failing spring of oxygen exists," continues Prof. Ellis, " and its copious streams, *by a nice adjustment*, replace by far the greater part of the loss. In the green grass, in the leaves of unpretending herbs and in those of the clustering woods, we shall find are hid those springs of this precious ingredient, *without which desolation and death might at no distant time gradually overwhelm the globe*." Prof. Ellis further states (folio 238) that the carbonic acid poisons which destroy animal life and provide sustenance for disease germs, are furnished to the air by various processes of combustion, respiration, putrifaction, and from volcanic craters, etc., which constitute the true source of vegetable nutrition. The composition—when in a normal condition—being, Carbon . . 27.27. Oxygen . . 72.73: " Each hundred parts of carbonic acid contains by weight 27¼ parts of carbon; if, therefore, we could remove the oxygen (which *we* have not a little done and still continue to do by our suicidal destruction of " clustering woods" all over the earth) " carbon is left." Yes, carbon, like the poor, " ye have always with you."

FOREST INFLUENCE IN GALLILEE.

" There was a time," declared an able writer in the *New York Herald* of July 12th, 1891, " when every acre of Gallilee not under pasturage was verdant with the foliage of trees. When the trees re-appear, as they might in a few years, Gallilee alone would be capable of maintaining an immense population in rich abundance." The above paragraph from an article on the probable return of the still scattered Jews to Palestine speaks volumes. Palestine, like all other parts of this earth, has been transformed into a barren destructive-insect and storm-breeding Sahara since the forest lungs were destroyed, and—deny it who may—the only possible remedy to restore the earth's atmospheric equilibrium there and elsewhere is to forthwith replant and properly protect our forests. On this momentous question I must now solicit the readers special attention.

THE ORDER OF CREATION.

The order of creation or evolution upon the globe shows clearly how very closely related every atom in and upon it is to each other. In each phase of being the manifestation of that period that then was was apparently independent of that which was to follow, but when that which was to follow made its appearance it found itself dependent upon everything which had preceded it or by which it was surrounded. Our dependence upon creatures in a lower stage of development admonished us to protect and nourish them in return—to " replenish the earth and subdue it." Surely, if words have any meaning at all, we cannot be blind to the obvious signification of the replenishing condition. What is replenish but to fill up again? If, according to the commonly received opinion among the churches, Adam and Eve were the first human beings placed upon the earth, it would be the height of absurdity to talk of them replenishing the globe by means of their offspring. You cannot fill again that which has never been full, but has always been empty before. If you set a savory dish and a clean plate before a guest you do not say to him, before he has touched the former, " Replenish your plate," but you may do so after he has filled it and eaten its contents. In that case we should be greatly surprised and vexed if, after satisfying his hunger, he should dash the plate to atoms on the floor and fling away the rest of the food out of the dish. Nevertheless, this is what we have all been doing for many centuries past, with respect to the earth, which was confided to our care to keep it and to dress it for each others unstinted happiness. Instead of replenishing it, instead of cultivating it in such a manner as that its fertility and productiveness should not only undergo no dimunition at our hands, but should go on increasing from age to age, we have defaced, deformed and devasted it. We have not been faithful stewards and honest husbandmen, but wasters and spoilers. We have desolated what we should have rendered still more fruitful and beautiful. In the impressive words of the Roman historian, " We have made a solitude and have called it peace." We have created a desert and charged the consequences of our own malignity upon God. Instead of causing the wilderness to blossom as the rose we have saturated the vineyards and orchards and harvest fields of some of the fairest portions of the globe with the red reign of ferocious warfare. We have felled the venerable forests, in whose green aisles myriads of winged choristers made blithest music from morn till noon, from noon till dewy eve; we have disrobed the mountains of their glorious atmospheric enriching foliage, and have transformed perennial springs into intermittent and terribly destructive torrents; we have exterminated whole species of quadrupeds of birds and fishes; we have introduced the utmost discord, derangement and disorder into the otherwise faultless harmony of nature; and we have torn from the bowels of the earth the precious metals which minister to our cupidity, and the minerals with which we decorate our persons in the spirit of the savage still surviving in us, leaving yawning carbon supplying shafts and caverns as pitfalls for all other comers. We have " smitten the earth with a curse," and that gentle, patient mother mourns and suffers by reason of the atrocious cruelty of her offspring, sprung from her womb and nourished every instant of our lives from her bosom, our crimes towards her are those of matricides. We never think for a single instant that if her bounty were suspended—if she were to cease to elaborate within her fruitful breasts the sustenance essential to the continuance of our vital functions, the whole of the human race would disappear and our globe would be as sterile as Sahara or the Polar ice, and " unless these days were shortened," some such a calamity would certainly occur, because never before in the history of the human race has the work of devastation proceeded with such frightful velocity and the consequent yearly increase of atmospheric troubles; never before has man been armed with such potent instruments of destruction; never before were these employed simultaneously in the five great divisions of the globe; never before had the restless spirit of " civilized" man and the various appliances by land and sea enabled him to penetrate into the heart of the African, the two American and the Australian continents, and to leave no island unpolluted by his defiling foot, no race of " savages" untainted by his deadly disorders. All the diseases with which the life upon and through the earth's surface teams, blight and murrain, plague, fire, flood, blizzard, earthquake and famine, all the unnatural disorders, confusion and turbulence of the elements, all the ravages attributable to draught and hurricane; all the contaminations of the atmosphere and the pollution of the streams and water-courses, are the work of the civilized races. We are reaping what we have sown, we are suffering the righteous penalties of our own misdeeds in earlier times. When we survey the melancholy ruins of cities that were once vast and populous, rising in dreary masses of shattered and unsightly masonry out of billowy hillocks of sand, in the midst of arid and sterile plains, and remember that these places were once surrounded by golden corn-fields and leafy groves, and gardens that were bright and fragrant with a tapestry of flowers and choicest of fruits encircled with umbrageous forests in which the deer browsed, and choirs of feathered songsters made music by day and night; and that noble rivers, which have entirely disappeared, wound their way in coils of glittering silver through grassy valleys, which afforded pasturage to countless herds of sheep and cattle, we might well shudder at the thought that the crime of having wrought this cruel transformation is chargeable, not upon a race which has passed away from or out of the globe, but upon *ourselves.* The generation is unchanged ! The same evil and destructive minds, grown more evil and destructive by reason of their maturity, are all here now, and they are repeating to-day in a much more vicious form the very acts of devastation which they performed in other parts one, two, three, four and five thousand years ago, when Jacob gathered his sons together in order that—instructed by God—he might tell them what should befall them "in the latter days." To what does any sane man imagine the Most High to refer to but the last incarnations of the twelve who were all gathered around Jesus 1800 years afterwards, and who are all in the flesh at this moment? For they had been with Him " from the beginning" of time, and will be so at the end.

"THAT WHICH HATH BEEN IS NOW."

Indeed the very identity of human actions in all ages, and in all countries should suffice to convince those who have any understanding that the same spirit of disobedience have been operating throughout. Once we ravaged and desolated a portion only of the surface of the globe; now we are making haste to ruin the whole. Look at what we have done by the wholesale denudation of forests all over the earth. Some of the great lakes of Europe and Asia are gradually drying up, because the loss by natural evaporation is no longer compensated for by the influx of tributary rivers; the volume of these being sensibly diminished by the destruction of woods in those regions in which they have their source. Thus, the level of the Caspean Sea, we learn, is 83 feet lower than that of the Sea of Azoff, and the surface of the Lake Aral is fast sinking. Twenty years ago it was shown that very large portions of the Ligurian province of Italy, *i. e.*, the Genoese territory, had been washed away or rendered incapable of cultivation in consequence of the felling of the woods. In Lombardy the demolition of the forests on the Appenines has led to the most disastrous results. The sirocco now prevails on the right bank of the Po, injuring the harvest, crops and vineyards; while the blasts of wind, which sweep across the country from the south and south-west, now assume the violence of hurricanes, and whole valleys are periodically visited with terrible inundations. In the district of Mugello all the mulberry trees have been destroyed with the exception of those which were indebted to neighboring buildings for a protection like that formerly afforded by the forests. North of the Alps we find the cultivation of the orange and many other fruits has had to be abandoned in certain situations on account of the late spring frosts, which were unknown until the mountain ranges had been stripped of their timber. Thirty years have sufficed to bring about these climatic disturbances in the French Department of Ardeche, as also in the plains of Alsace, where a much more genial temperature prevailed, as in the United States of America and Australia, before the neighboring forests were cut down. In Sweeden it has been observed that the spring commences a fortnight later in these districts in which the woods had been demolished than it did in the last century.

France has suffered to an immense extent by the de-foresting process. In one department alone, that of La Brienne, 200,000 acres which were once covered with woods interspersed with pastures are now bare of timber, and have been converted into a dreary and malarious expanse of pools and marshes. The same thing has happened on a larger scale in La Sologne, where as much as 1,000,000 acres of land that were well wooded, well drained, and productive, are now barren and desolate. This deplorable state of things is explained by a fact familiar to every forester, namely, that, trees are instruments of drainage. Their roots often pierce through subsoil almost and in many instances quite impervious to water, and in such cases the moisture which would otherwise remain above the subsoil and convert the surface earth into a bog, follows the roots downwards into more porus strata, or is received by subterranean canals or reservoirs. When the forest is felled the roots perish and decay, the orifices opened by them are soon obstructed and the water having saturated the vegetable earth-mould stagnates and transforms it into ponds and disease-germ breeding morasses.

In M. Marchand's excellent work, entitled *"Les Torrents des Alpes et le Paturage"* is the following passage descriptive of what is now going on in the civilized world: " Unhappily, man, improvident and avaricious, has frequently destroyed the forests that he may thereby get possession of the soil. He has substituted for them pasture grounds, often ill maintained. With the ruin of the soil begins that of the people. The more unhappy they are, the more selfish do they become (and the converse of the proposition holds equally good) and the more they destroy, so that from the time evil begins it cannot but go on increasing."

BOUSSINGAULT.

Boussingault speaks of the two lakes near Ubate in New Grenada, which a century ago formed but one. When he visited them he found the waters gradually retiring, and vegetation encroaching on the abandoned bed. The enquiries which he instituted, satisfied him that the circumstances were attributable to the extensive clearings which were going on all around it. In the same valley he ascertained that the length of the lake Fuquene had been reduced in five centuries, from ten leagues to one and a half, and its breadth from three leagues to one. At the former period the neighboring mountains were well wooded, but at the time of his visit they had been almost entirely stripped.

HUMBOLDT.

That close observer, Alexander Von Humboldt, noticed the same thing in regard to the Lake of Valencia which had been diminished from year to year because the loss by evaporation is not made good by precipitation. So rapidly had this been proceeding that some people imagined the lake must have a subteranean outlet; but Humboldt clearly perceived and has lucidly explained the cause: " By felling the trees which cover the tops and sides of mountains—he observes—" men in every climate prepare at once, two calamities for future generations, want of fuel and scarcity of water. Trees by the nature of their perspiration and the radiation from their leaves in a sky without clouds—as in the regions of which he was writing—surround themselves with an atmosphere constantly cold and misty. They affect the copiousness of springs, not as was long believed by a peculiar attraction for the vapors diffused through the air, but because by sheltering the soil from the direct action of the sun they diminish the evaporation of water produced by rain. When forests are destroyed as they are everywhere in America by European planters, with imprudent precipitancy, the springs are entirely dried up, or become less abundant. The beds of the rivers remaining dry during a part of the year, are converted into torrents wherever great rains fall on the heights. Hence, it results that the clearing of forests, the want of permanent springs and the existence of torrents are three phenomena closely connected together." Humboldt might safely have added another and equally serious phenomena namely, a foul destructive insect pest breeding atmosphere.

REBOISEMENT IN FRANCE.

Dr. Brown's valuable work on " Reboisement in France," which contains the essence of many previous books on the same subject, and notably that of M. Surell's " *Etude sur les Torrents des Hautes Alpes*," deals exclusively with the causes, consequences and correctives of the Alpine torrents; proves that man is responsible for them, exhibits their appalling effects, and describes what measures have been or ought to be taken, to check the growth of this terrible evil. Dr. Brown, page 46, alludes to the torrent producing cause as follows: " Seeing then a very remarkable double fact; everywhere where there are recent torrents, there are no more forests; and wherever the soil has been stripped of wood, recent torrents have been formed; so that the same eyes which have seen the forests felled on the slope of a mountain, have there seen incontinently a multitude of torrents.   The whole population of this country (the sub-Alpine region) may be summoned to bear testimony to these remarks.   There is not a commune where one may not hear from old men, that on such a hillside now naked and devoured by the waters there have been formerly fine forests standing, without a torrent."   Nothing can be simpler than the relationship of cause and effect in this case.

A FEW OF THE FRUITS FROM WHOLESALE FOREST DESTRUCTION.

A little over two years ago the following appalling particulars appeared under the heading of " The Unfortunate States" in the London *Evening Standard*, and reappeared in the Melbourne (Australia) *Daily Telegraph* of September 20th, 1890 :—

" The United States are at present suffering from a varied series of calamities. During the last few months cyclones have again and again swept over broad acres of the country, and now within the space of four and twenty hours earthquakes, rainstorms and tornadoes have been doing their worst to make many parts of the country, from New York to the Rocky Mountains, less endurable than ever. . . . . The torrents of rain commenced, no doubt, with the tornadoes, which have, however, been more destructive, while the latter have not, it seems, extended over so wide an extent as usual.   The most aggravating feature about these great wind-storms is that *by no possible contrivance can they be either lessened in violence or their effects diminished by one iota*.   On the contrary, *as time advances and the country gets more thickly settled* (on *deforested soil*) *their destructiveness must necessarily become greater* (the italics are mine).   All that science is likely to ascertain regarding their origin, nature and progress, have been already garnered into ample repertories of American meterology, and though, thanks to the recent researches of Farrell, Davis and Hogan, the *theory* of the cyclone and tornado is now almost perfect.   This perfection affords no hope of the future bringing any relief to the sorely-tried dwellers in the western States. The many peculiarities of the American climate are due to the unique position of the new world, and especially of the United States, in being placed between two oceans, and bounded on the south by a tropical sea like the Gulf of Mexico, and on the north by the eternal ice of the Polar ocean.   Its breadth is also productive of inconvenient consequences, thus while the Atlantic and Pacific sides of the Continent are, so far as they can be reached by the moist breezes from either ocean in the enjoyment of a plentiful rainfall, the central region of the prairies is dry and arid for the greater part of the year, extremely cold in winter, and more than usually warm in summer.   The latter circumstance, to a large extent, accounts for this section of the United States being the scene of the violent rainstorms and whirlwinds which have *of late years* attracted more and more attention *owing to their frequency* and the loss they cause to life and property.   Earthquakes, though by no means rare in the Mississippi valley, are neither so numerous or so violent as in California.   Whirlwinds are common features of every dry hot region, being

due to a disturbance in the equilibrium of the atmosphere by the excessive rare-
fraction of the air at one particular spot with the sudden inrush from several directions to fill up the vacuity thus caused ; of these the sand-pillars of the Asiatic and
African deserts are the most striking concomitants, while at sea the water spouts
form the equivalent of the sand pillars which are the materialized bases of the
"afreets" of the Oriental mythology upon the land.

"The revolving of circular storms variously known as cyclones, tornadoes and
hurricanes are in some respect different from these. Indeed, though a cyclone
and a tornado are in ordinary parlance considered synonymous terms, they are
metereologically distinguished by various points, special to each of them. Thus,
though a cyclone is generally understood to be a storm of extraordinary violence,
it must be remembered that the gentlest inflow of the air to fill up a vacuum is in
kind, if not in degree, identical with the wildest hurricane which levels everything
in its tract, but still they differ in some minor characteristics. Thus the great
circular storms which destroy such enormous masses of property, and often so
many lives, are confined to certain areas, none of which are on the equator *as yet.*
In America, the Mississippi valley particularly, the broad strip from western
Ohio to Colorado is the district which suffers most," (every other State now, 1893,
more or less participates in the climatic disturbances) " and though not confined
to any period of the year, the storms are most frequent in April, May, June and
July, and on the afternoons of the days rendered memorable by their occurrence.
The ordinary whirlwind is to be traced to a thin layer of heat, and therefore at-
tenuated air next the ground, which is, however, of too small an extent to pre-
cipitate any powerful movement of the atmosphere in its upward whirling. The
tornado, which on the other hand, is due (so Farrell, the latest of the tornado in-
vestigators, says) to a thick layer of hot, moist air between the earth and the
denser upper atmosphere in rising is cooled by expansion, and the invisible vapor
with which it is laden condensed, thus liberating a large amount of latent heat."
(Only from deforested and prairie lands). "This latent heat still further rarefies
the current of ascending air rushing in on all sides to fill up the vacuum thus created. In this manner the layer of heated air, owing to its great thickness, causes
in its upward movement a correspondingly violent disturbance of the atmospheric
equilibrium." (Blind indeed must we be to overlook the fact that a terribly destructive "thick layer of hot, moist air between the earth and the denser upper
atmosphere," could not exist were the earth's forest lungs thoroughly restored and
properly sustained in and around those severely afflicted parts, as the primary hot
moist air creating cause would then be greatly diminished besides providing abundant leafy absorbing reservoirs to profitably entertain all such vacuum forming
air). " But," continues the writer, " whether the storm is the comprehensive
cyclone or the narrower but fiercer variety of it, known as the tornado, it is difficult to differentiate between the damage done by the one and that which follows
in the train of the other. Little warning is accorded the ill-fated settler of the approach of the winged monster whose appearance drives away every thought save
that of flight *to some underground cavern"*—a nice prospect for agriculturists—
"since few ordinary buildings are strong enough to resist the full impetus of such
a wind. Dark and threatening clouds appear in the west, and a lurid or greenish
tinge suffuses the sky in the same quarter towards the south, clouds of dust soon
follow, concealing the funnel-shaped cloud in the rear, then, as the tornado nears,
an indescribable loud roar is heard. The bellowing of a million of mad bulls and
the roar of ten thousand trains have been among the similes suggested, though,
perhaps, a continuous roar or rumble of thunder may best describe this dismal
forerunner of the storm. But the observer has little time for reverie. In a few
moments the funnel-shaped cloud itself follows like a great balloon, sweeping the
neck round and round with terrible fury and destroying everything in its track.
In three or four minutes it has passed by, but in that short space of time the
staunchest houses of brick and stone have been demolished, and sorrow and ruin
spread all along the path. Indeed, so narrow is this track that in the great tornado of last March (1890) the wind in many instances mowed a swath for itself,
levelling every building or object in the direct route, but leaving whole or partly
uninjured those on either side of the mad rush. Men, women and children,
sheep, cattle, pigs and horses are carried off their feet, and even borne on the

wings of the wind for a considerable distance in spite of their lying prone on the ground or clinging to what seemed stationary objects; railway trains were overturned and swept off the line; steamboats sunk in the rivers, and large wooden structures lifted up bodily and carried off several yards from the place where a minute earlier they appeared as if immovable. Owing to the sudden decrease in the pressure of the outer air, the atmosphere within the walls of the house in the course of the storm, acts against the sides of the structure, so that buildings which might otherwise brave the blast are exploded as if a spark had been applied to a roomfull of gunpowder. Fowls have been known to be stripped of their feathers, little streams emptied of their water, heavy iron chains blown through the air, nails driven head-first into planks, stalks of Indian corn shot through a door, large beams tossed with such impetus that they penetrated the earth a foot or more, and in India" (where similar visitations are experienced from a similar cause) "a case is on record in which a bamboo was actually propelled through a mud wall five feet thick with a force equal to that of a cannon discharging a 6 ℔ ball."

ALARMING DISPATCHES TO THE S. F. EXAMINER, 1893.

ATLANTA, (Ga.), March 4.—Georgia was visited by a cyclone last night the reports from which indicate great loss of life and immense destruction of property. The town of Greenville, having 1,000 inhabitants, was swept out of existence with, however, the loss of but one life. A small town called The Rock, a few miles off, fared worse, as five lives were lost there. Near Barnesville the cyclone dipped to the ground again, and three more people were killed. In East Mississippi the storm seemed to have done great damage, completely wiping out the three towns and wounding and killing many people. The cyclone, after sweeping across Mississippi and Alabama, struck Georgia at a point on the Chattahoochee river below Columbus and divided into two sections, one following the course of the Chattahoochee, going north of Atlanta, passing over Rome and on through the Blue Ridge mountains into North Carolina ; the other branch pursued a course across the State south of Macon, passing on north of Augusta and through South Carolina, where it united with the northern section, and passing through Wilmington, N. C., found its way out into the ocean—(to destroy the shipping).

It was a few minutes after 8 o'clock last night when it struck Greenville. The first building to give way was the Courthouse, which was blown to atoms. In almost an instant the buildings generally began to grate and fall from their foundations. The people, thoroughly affrighted, could do nothing in the wreck and confusion which surrounded them. The night was intensely dark and the weather was bitterly cold. The storm lasted but a few minutes, and when it was over the people found themselves without shelter, and had to go to work to improvise such covering as they could. The fear that a great many had lost their lives added to the terror of the occasion. Investigation developed the fact that but one person, a negro, had been killed, though a great many of them were wounded, some of them severely. The Presbyterian church, the postoffice and the college were blown to atoms. The people are in great distress and all are without homes. The town is one mass of ruins, and the damage is beyond description. The next town in the path was Hogansville, where several houses were lifted up and carried two miles where they were dashed to the ground. The Rock, a hamlet of 500 inhabitants in Pike county, suffered a loss of five lives. Among those killed was Judge Riviere, a prominent citizen of the county. At a point about ten miles west of Barnesville three deaths resulted, but it is impossible to get their names. In Le Grange four houses were blown down. Mrs. Ross, a lady who lives near Piermont, lost her life, and many others were badly injured, some fatally. The storm's course was down the Atlanta and Florida railway toward Barnesville, and at Pine Mountain, in the neighborhood around Barnesville, the following deaths and casualties are reported : Miss Daisy Hawkins and one unknown colored man killed; near Piedmont, two colored children killed. There are also numerous names being furnished of persons who have been injured.

From all along the path of the storm come reports of the loss of life, which must run the list up to about fifty. The general course of the storm lay across the country out of the lines of railway travel and with very little telegraphic communication. It followed the course of what is known as the Harris county track. Many of these cyclones have kept as closely within the lines of their predecessors as if their course had been laid out by an engineer.

The maximum wind velocity in Atlanta to-day was forty-four miles an hour. The hardest blow here last night was thirty-six miles an hour. The wind was blowing thirty-six miles an hour at Savannah and twenty-eight an hour at Augusta at 7 o'clock this morning. The Weather Bureau synopsis to-day shows that the trough of low barometer pressure, which yesterday morning extended from Texas northeastward to the New England States, developed into a storm of considerable area and energy, and enlarged toward the Atlantic coast until it covered the entire country east of the Mississippi river, its center being near Wilmington, N. C., with a minimum pressure of 29.30 inches.

MERIDIAN (Miss.). March 4.—The havoc wrought by the cyclone in this section last night is incalculable. The scene at Marion, Miss., is one of awful destruction. The main track of the storm was 300 yards wide, and everything in its path was swept away, the wreckage of houses being scattered for miles along its course. The cyclone struck only the northern portion of the town, which is but sparsely populated. The injured are J. Harrison and wife, George Nailers and Mrs. White. Mrs. Meader and her daughter were killed. Half a mile of telegraph poles were blown down. Four settlements of negroes were destroyed, but no one seriously injured. The town of Toomsuba was almost completely wrecked and a number of people injured. At Keating a negro settlement was almost completely destroyed.

SPECIAL DISPATCHES TO THE S. F. CHRONICLE.

MEMPHIS, (Tenn.), March 23, 1893.—The most destructive cyclone in the history of this section swept over Northern Mississippi and Western Tennessee late this afternoon, leaving death and destruction in its wake. Kelley, Miss., a town of about three hundred inhabitants, was wiped off the face of the earth, every building in the place being totally demolished. So far as is known twenty-five people were killed outright and about sixty injured. The cyclone reached Kelley about 3:40 o'clock this afternoon, spreading havoc in every direction. Long before the wind struck the town a strange atmospheric condition was noticed. The air grew very dark and then a moaning sound was heard, and finally a greenish colored cloud was seen rapidly approaching from the southwest.

The path of the storm was about half a mile wide, and everything in its course was picked up like a straw and dashed to pieces. Large houses were crushed like eggshells, while giant forest trees were uprooted and their trunks were picked up by the whirling wind and carried for miles. The public school building was the first to go down before the storm. The pupils had been dismissed only a few minutes before and most of them had left the building, which fact alone prevented appalling loss of life. Several children were caught in the ruins, however, and crushed to death.

A row of frame buildings next fell before the cyclone's fury, and with a loud crash and a deafening roar they were literally torn to kindling wood and the fragments scattered far and wide. Owing to the darkness it is impossible to learn the full extent of the loss of life and property. Trains from the East reaching this city late in the afternoon and evening brought reports of widespread destruction. The passengers on a Yazoo and Mississippi Valley train told of the destruction of Tunica, Miss., and, while the reports were slightly exaggerated, yet they were in a large measure confirmed. Tunica suffered greatly, but the loss so far as is known was confined solely to property. A special from there says the damage to property will run into hundreds of thousands of dollars.

About 3:30 P. M. the sky in the southwest began to darken and a low wailing sound announced the storm coming. Within a few minutes the wind came along with terrible velocity and with a swish and whirl that portended danger. The first hard blow gave way to the cyclone and houses were crushed like eggshells. The vicious visitor lingered over and around the town for scarcely two minutes, and yet in that time it leveled buildings unsparingly, tossing saloon and church alike to the ground. Such an unusual and unexpected visitation stunned people and the noise of tumbling roofs paralyzed their minds for the moment.

A partial calm, save for the fall of heavy rain, then came and the people rushed about in great excitement. On one side of the square, where stood a handsome building occupied by the Knights of Pythias and Masons, were now only a heap of timber and beams. This was one of the most pretentious buildings in town. The people on the streets first noticed this wreck, and then they saw that the roof of the courthouse was gone.

But there was more than this; there were the cries and screams of children. Men rushed to the colored schoolhouse where 150 children had been gathered at their lessons. The building, a two-story frame, had been blown down and beneath the ruins was a mass of struggling children. No lives were lost, but there were many maimed and crushed, some with broken arms and some with fractured ribs. The work of relief was at once directed to the schoolhouse, and the children were extricated from the prison which the timbers formed. The full list of the buildings wrecked cannot be obtained to-night, but it is known that the Presbyterian and Methodist churches are totally wrecked. In all parts of the town are piles of ruins, and very few houses have escaped without some damage.

The path of the cyclone appears to have been twenty miles in width, although the serious damages was confined to a much smaller area. The wires are down in all directions. There is no telegraphic communication whatever with Nashville and intervening points, and very little news is obtainable from the places visited by the cyclone. This city barely escaped. A heavy rain fell and a high wind blew at the time the cyclone raged, and it became as dark as night for thirty minutes, but no damage was done. The train from Birmingham, Ala., arrived several hours late and reported much damage between here and Bybalia, thirty miles east. Farmhouses, barns and ginhouses are reported unroofed and blown down all along the line. The train was delayed by having to stop at frequent intervals to chop away large trees that had been uprooted and blown across the track. Damage is reported at Captville, Tenn., and Olive Branch, Miss., but no particulars are obtainable.

MEMPHIS, (Tenn.), March 24, 1893.—The damage done by yesterday's cyclone in the Mississippi valley is enormous. While the loss of life is not as great as was at first reported, the damage to property will reach $2,000,000. The telegraph wires are still demoralized and reports are coming in slowly from the storm districts. It will be several days before the full extent of the disaster will be known. The death list at 10 P. M. foots up twenty-three, while the list of the injured will run up into the hundreds. The names of the dead at Kelly, Miss., so far as known, are: Harriet Smith, Mary Williams, Susan Williams, and two unknown negro women. The dead elsewhere are: Richard Heard and Thomas Heard, Shubuta, Miss.; Eli Prince, Evansville, Miss.; Drury Sumralls and his family of nine, Shaw's, Miss. The names of the injured at Kelly, so far as known, are as follows: Richard Pine, wife and children, all badly injured by the collapse of their house, one fatally; Jim Payne, so badly wounded about the head and shoulders that he may die; Chris. Burford, internally injured and will probably die; Mrs. Sarah Hart, two ribs broken and internally injured, may die; Marion Mason, cut about the head; Mrs. Mason, badly hurt about the hips; Harriet Branch, internally injured; Gus Bills, right eye knocked out; Eph. McLaughlin, shoulder broken; F. Wiley McLaughlin, arm dislocated. The injured at other points are: S. K. Davis, Crawfordsville, Ark.; fourteen negro tenants at Crawfordsville, Ark., more or less seriously injured; John Carroll, Spring Creek, Tenn., struck by flying timber and seriously injured; twenty-one school-children at Tunica, Miss., more or less seriously injured. The majority of those killed and injured are negroes.

The first heard of the cyclone was in Northern Lousiana and Southern Arkansas. It crossed the Mississippi a few miles above Greenville, devastating plantations, wrecking farmhouses and uprooting giant forest trees. The path of the storm was about half a mile wide, and nothing was left standing in its track. The first fatality occured near Shaw's station, Miss., where the house of Drury Sumrall, a prosperous colored farmer, was leveled to the ground, killing the entire family of nine persons. The cyclone passed through the suburbs of Shaw's and demolished several residences and small stores, but no one was killed. The hurricane then changed its course slightly and traveled along the right of way of the Yazoo and Mississippi Valley Railroad until it entered Cleveland, Miss., where a public school building and several stores and residences were raised to the ground. No fatalities occurred at Cleveland, but several people were struck by flying timbers and more or less injured.

Leaving Cleveland the cyclone passed within a mile of Clarksdale, a town of 2000 inhabitants, and next struck Tunica, the county seat of Tunica county. Nearly every building in the place was wrecked. The recently completed courthouse went down before the wind. The colored school building was wrecked and over thirty children were maimed and crippled, some of them being fatally injured. As the cyclone left Tunica it divided, one portion traveling in a northeasterly direction while the other took a northwesterly course and again crossed the Mississippi into Arkansas, where it spread ruin in three counties. The towns of Crawfordsville and Vincent were nearly wiped off the earth. The storm then took a northeasterly course, reaching Kelly, Miss., about 4 o'clock in the afternoon. Here the greatest damage was done. Six people were killed outright and scores were injured. Not a building was left standing, the fragments being strewn over the country for miles. Physicians from the surrounding towns hurried to the scene and rendered all possible aid to the sufferers. Temporary hospitals were fitted up in farmhouses that had escaped the storm, and those who were fortunate enough to get out uninjured did all in their power to help the sufferers. The damage to property in the vicinity of Kelly will reach $150,000. After leaving Kelly the cyclone passed into Tennessee, the next place visited being Spring Creek, a small town in Madison county, where several people were injured, but no one was killed. No reports of damage have been received beyond Spring Creek, except in a suburb of Nashville. The path of the storm after it left Madison county was through a country remote from telegraph lines, and it will be several days before full details will reach the outside world.

MOBILLE, (Ala.), March 24.—Early this morning a cyclone passed one mile north of Shubuta, Miss., going southwest. At Arista John's place a tenement house containing negroes was leveled and two negroes were killed and three were injured. One mile east of there houses were blown down. Ten miles further east three tenement houses were destroyed, but no one was hurt.

NASHVILLE, (Tenn.), March 24.—One of the most terrific wind and rain storms in the history of Nashville swept over this city last night. The greatest force of the storm was felt in the northern part of the city, where several houses were unroofed. One occupied by W. F. Bradford was completely razed to the ground. Bradford was taken from the ruins in a badly bruised condition. McNeil Drumright, aged 13, who boarded with Bradford, was taken from the debris in a mangled condition and cannot live. Eugene Drumright, aged 18, a brother of McNeil, was horribly mangled and dead when found. It is feared others were injured or killed in the building. In all fifteen houses were badly damaged or destroyed and the loss will foot up in the thousands.

LOUISVILLE (Ky.), March 24.—The heavy storm which passed over a large section of the South last night did great damage at Bowling Green and the surrounding country. The roofs of fifteen or twenty houses were blown off, the most serious damage being done to the Louisville and Nashville roundhouses, which were unroofed and the walls leveled with the ground. Falling material did serious damage to engines inside. One colored man was caught in the debris and badly, though not seriously injured. The loss on the building and locomotives is from $75,000 to $100,000. Passengers on the delayed fast express on the Louisville and Nashville from the South stated that all along the road evidence of the storm could be seen. Many farmhouses were roofless, and scores of stables and outhouses were totally demolished. The town of Rowlins was almost destroyed, the postoffice building being swept entirely away, while damage to others was very heavy. Every house in Stonford was damaged. At Murray, Ky., twenty residences and fifty stables and barns were demolished. Only one person, Miss Aline Stabblefield, was seriously injured. A dozen were slightly hurt.

INDIANAPOLIS, (Ind.), March 24.—A cyclonic storm visited Indiana last night. In this city fifty houses were wrecked in one neighborhood in the northwest portion and many families are temporarily homeless. At Tuxedo, a suburb, many houses are wrecked and several small ones were carried away. The Capital City Coffin Works are badly damaged. Advices from all parts of the State indicate that much property is damaged and some persons were maimed. At Loogootee the flouring mill, the Catholic church and City Hotel were badly damaged. At Evansville the south wing of the insane asylum was damaged. At McCordsville the house of James McCord was blown down and Mrs. McCord was fatally hurt. At Brazil outbuildings, fences and trees were leveled and coal mines were flooded. At Alexandria residences and business blocks were damaged, and the Lippincott glass factory was destroyed, John Angle, Jr., being instantly killed. F. McShafery, Peter Hanlan, Ernest Frey, James Branham and others were seriously injured. At Vincennes houses, barns, trees and fences were laid low for twelve miles. Several thousand dollars worth of property was destroyed.

And just for a moment pause to consider the following dreadful supplementary pictures clipped from the S. F. *Examiner* of April 8th, 9th and 10th, 1893:—

A TERRIBLE STORM.

"DEADWOOD, (S. D.), April 7.—A terrible wind and snow storm has been prevailing here for the past forty-eight hours. Telegraph and telephone wires have been prostrated, many buildings blown down and others unroofed. Piedmont has been partially destroyed. All the trains have been tied up. The velocity of the wind is seventy-five miles per hour."

BUFFALO, April 8.—One of the worst cyclones that ever swept western New York struck this end of the State yesterday. Reports have been coming in all day of damage done in Chautauqua and Erie counties. As near as can be gathered the hurricane struck near Springville, in this county, and then swept down across Chautauqua Lake and into Lake Erie. A dispatch from Springville states that a barn belonging to Vedder Hemstreet was blown down. Mr. Hemstreet and his hired man were in the barn with some cattle when it collapsed. The latter was caught between two cows and escaped unhurt, though the cattle were crushed. Hemstreet was pinned between several timbers and died before he could be extricated.

At Brocton, the heart of the grape country, tremendous ravages in orchards and vineyards are reported. Trees where uprooted and buildings leveled. At Westfield trees were uprooted and a water tower and windmill were lifted bodily and moved a couple of yards. The Lake Shore and Michigan Southern tracks between Angola and Farnham were washed out. Fair buildings at Dunkirk were demolished. Many cattle in Sheridan were killed. Churches, printing offices and other buildings in Fredonia were stripped of their roofs. At Mayville a boathouse was scattered over a farm and Lake Chautauqua was lashed into fury. At the Assembly grounds several handsome trees were shattered and broken and cottages twisted from their foundation. In all these towns narrow escapes from death are reported.

The residence of George H. Talcott, at Talcottville, Lewis county, was struck by lightning some time during last night and burned to the ground. Talcott and his brother were burned in the house, their charred and blackened corpses being found in the ruins this morning. The damage done to the buildings by the storm cannot be less than $100,000. It is impossible at this writing to estimate the injury to orchards and vineyards.

CURSED BY CHOLERA.

[Special to the EXAMINER.]

LONDON, April 8.—If last year's devastations of the scourge were not fresh in the public mind, Europe would be already in a cholera panic. The disease probably exists to-day in a larger number of towns than when the epidemic was at its height in Hamburg last summer. The criminal policy of concealment is again being pursued in many places. The most outrageous are in towns on the northern coast of France, where it is known that nearly a hundred deaths have occurred within a fortnight. Russia acknowledges several hundred deaths over her vast extent of territory, and it can only be guessed how much this is short of the truth.

Strong appeals in the advertising columns of the St. Petersburg newspapers for the services of doctors in infected districts indicate how great is the emergency. Not a single medical or scientific advantage is yet announced as the result of the cholera conference at Dresden. The object there sought has been solely how to mitigate the commercial evils of the epidemic.

FEARFUL SHIPWRECKS.

" Over in Hydrographic Officer Burnett's corner of the Merchants' Exchange is a chart that seems to have been drawn by somebody who had a nightmare of meandering red streaks, little red and blue dots and crosses, and red and blue representations of the bows of partially submerged vessels. It is a wreck chart showing the vagarious driftings of the derelicts on the North Atlantic during the past five years and to the mind of a nautical man is the publication *chef d'oeuvre* of the office since Commander Glover assumed charge. The best statistics obtainable show an average annual loss of 2,172 vessels and 12,000 lives in the marine commerce of the world, entailing monetary losses amounting to $100,000,000. In the period of five years covered by the chart there were 956 vessels wrecked on the Atlantic coast, and in the same time 1,096 vessels were abandoned and left to drift about the ocean and menace the safety of navigators whose courses might lay in the way of the drifters. Of this thousand and odd craft there were 625 unknown; 332 have been located once or more, and 139 have been sighted so often as to permit of their drift tracks being charted."

FIERCE FOREST FIRES.

[Special to the EXAMINER.]

CINCINNATI, April 9.—Dispatches from Vanceboro, Lewis county, Ky., on the Ohio river, seventy-five miles from Cincinnati, says that fires in the forests in that country broke out several days ago and have spread over the whole country. Last night, from Clarksville, in Lewis county, to Sugar Loaf mountain, the whole country was one vast sea of flames. Fences have been destroyed everywhere, and the houses of many farmers have been burned. A dispatch from Chillicothe says that extensive fires are raging in the hill forests near Bainbridge, Ross county, and are spreading in the hills of Pike and Highland counties, near by, doing great damage.

PORTSMOUTH (O.), April 9.—For the past two weeks very strong and dangerous forest fires have prevailed west of the Scoto river. The hamlets of Union Mills and Friendship were surrounded by fire, but rain saved them. At present the fire is smoldering and another rain will quench it. The loss, it is estimated, will exceed $200,000 in timber, etc., that was burned, not counting a score of farm buildings swept away.

I have thus fully quoted a few of many such like newspaper reports relating to awfully disastrous atmospheric troubles which of late years have caused wide-spread ruin through very many portions of this great country and on the sea, the former to show the special necessity which really exists for a whole-hearted restoration and preservation of forests because of "the unique position of the new world" so graphically detailed by the writer, and the others to illustrate the dreadfully savage and increasingly erratic nature of recent storms which compel long-suffering agriculturists and others to burrow through underground caverns so as to escape being swept away and destroyed with their homesteads ! Surely such atmospheric conditions are the very reverse of normal?

Professor J. R. Buchanan, M. D., of New York, who published a book some years ago pleading for the establishment of ethical and industrial education, and who has further published a manual of psychometry in which he made several predictions, which have been verified, published in the *Arena* for August, 1890, a most alarming paper, entitled, "The Coming Cataclysm of America and Europe," leading up to terrible devastating torrential storms, earthquakes and almost general barrenness, extending over vast regions, making special mention of American and continental deforestation as a primary cause.

FOREST DESTROYING COMBINES.

And this is how, **for many years, forests have disappeared in** addition to the selectors clearings **in America.** I quote from *Harper's Weekly* for July 18th, 1891 :

"The condition of lumber mills and saw mills in Michigan, Wisconsin and Minnesota, 1890, has been announced. Only planing mills operated by lumber manufacturers in connection with lumber mills include only those which manufacture sawed lumber as the principle product, the term "saw-mills" meaning all other mills in which logs or bolts form the principle raw-material, and are manufactured into any kind of product other than lumber. The value of forest products, not manufactured at the mill, in Michigan, Wisconsin and Minnesota, 1890, aggregates $30,426 194; value of mill products $115,699 004; value of re-manufactures $21,112.618—making **an** aggregate value of products in three States of $167,237.816. The capital invested to produce this value was $270,152 012. Men employed in forests, 95,258; women, 99; children, 10; animals, 32,491. In the mills, **the** product required the labor of 87.939 **men, 646** women, and 653 children. The amount represented in **operation of machinery** and chemical appliances, **1890, was** $23.559.334; the **expenditure of steam and water** power was reported **as** sufficient to lift 3,500,000 tons one foot **in one** minute; 1,262,151,180 *cubic feet of merchantable timber were removed from natural growth;* $7,890.254 **were invested** in vessels **and** other means of transport, and $99.688.256 were **expended** for **wages.** The aggregate increase of product since 1880 is reported to **be** 29.66 per cent. in quantity and 57.92 per cent. in value."

In the three States named above there are 933 establishments—operating mills, with a capital invested in timbered land of $85,381.446, the area being 6.818,941 acres, with an estimated total product of merchantable timber of 43,133,886,209 feet (board measure). The estimated value of standing timber owned by these establishments is $135,612,007. White pine is by **far** the most important product, the total on timber land is estimated to be 47,304, 557,519 feet.

TIMBER PRODUCTS—The aggregate number of establishments engaged in the " timber product" industry in Michigan, Wisconsin and Minnesota is 574, with a capital of $46 765,405. **The** product of **logs for** domestic manufacture, 1890, is shown as 1,392.585,874 **feet,** that of **hard wood and** other logs for export being 33,115,000 feet."

When **we add to the above an** estimate of the annual demolition of timber in Oregon and other States all over the Union, as also in Canada, for general and now for paper making purposes, coupled with wholesale forest fires caused in nearly every instance from criminal carelessness, **we** should not very much wonder at the dreadful results now upon us. **But for the** rapidly increasing atmospheric penalties we would—as they do **in** " **treeless Spain**," continue to annihilate every health inspiring tree **for a** temporary selfish gain, **and** then ineffectually **strive** to extirpate the many consequent destructive **insect** plagues which, **as** "nature's scavengers" **are evolved and** thrive in the **poisonous** atmosphere which our tree killing **conduct creates.** The Italian poet, " **Dante**" when penning his " **Inferno**" early in the 14th century, **was evidently** permitted to forsee the outcome **of our** heartless ingratitude towards our revengeless tree friends' in these latter days, when he **wrote the following lines:—**

DANTE'S INFERNO.

" E'er Nessus yet had **reached the** other bank
We entered on a forest, **where** no track
Of steps had worn away, **Not** verdant there
The foliage, but of dusky hue; not light
The boughs and tapering, but with knares deform'd
And matted thick: *fruits there were none, but thorns*

Instead with venom filled. Less sharp than these
Less intricate the breakes, *wherein abide*
Those animals (insect pests) that hate the cultured fields

.   .   .   .
.   .   .   .   .   On all sides
I heard sad plainings breathe, and none could see
From whom they might have issued.   In amaze
Fast bound I stood.   He, (the inward guide) as it seem'd believed
That I had thought so many voices came
From some amid those thickets close conceal'd,
And thus his speech resumed: "If thou lop off
A single twig from one of those ill plants,
"The thought thou hast conceived shall vanish quite"
   Thereat a little stretching forth my hand,
From a great wilding gather'd I a branch,
And straight the trunk exclaim'd: "Why pluck'st thou me?"
Then as the dark blood trickled down its side,
These words it added: "Wherefore tear'st me thus?
Is there no touch of mercy in thy breast?
Men once were we that now are rooted here,
Thy hand might well have spared us, had we been ——
The souls of serpents."   (* PSALM 139, 15-16)

FOREST LAND PREFERRED FOR SETTLEMENT.

As in the United States and Canada so in Australia, forest lands were preferred by selectors for farming on, and after our thirty years of reckless toil, many millions of acres once clad with stately malaria destroying eucalypti whose gorgeous atmosphere purifying and rain attracting foliage formed a beautiful canopy of variegated colors under and around which, weeds and insect pests could not exist, now form very many unsightly, because parched weed and insect breeding fields, made hideous alike to the misguided owners and leasees as also to the weary traveller throughout the " settled" districts, by being compelled to gaze on monotonous forests of weather bleeched dead tree skeletons, left standing as so many reproving spectres on large areas of sun baked agricultural and pastoral lands.

New South Wales—the oldest Australian colony, originally known as " the Botany Bay convict settlement" has provided ample insect plague breeding fields for the whole group.   Ticket of leave convicts and crown pastoralists having added small orchards and vineyards to their respective homestead belongings which in course of time from various avoidable reasons, were permitted to run wild in exhausted soil, thereby furnishing the necessary insect breeding conditions, which eventually charged the passing winds with destructive germs.   The following dispatch clipped from the *Sydney Daily Telegraph* of February 4th last, faintly illustrates some of the results:—

ADELAIDE, Friday.—The Crown Lands Department reports that the ravages of the codlin moth, rather than diminishing, are spreading amongst the apple orchards.   Spraying with Paris green has been resorted to, without satisfactory results.   There are now more infected orchards than last season, and the damage to fruit has been enormous.   There are at present 169 orchards affected in the colony.

Whilst the forests remained in tact, through neighboring colonies, destructive insect raids from said fields were however comparatively harmless. Early in 1870, about eight years subsequent to the first general land selecting scramble for agricultural settlement in the colony of Victoria, I observed myriads of minute little grasshoppers like so many sand ripples, in a remote arid portion of said colony adjacent to the New South Wales border line, slowly but steadily making their way towards the settled districts, and although the leading authorities were duly notified of the invasion, little or no heed was taken of the warning—the supposition being that they would " perish on their way south." In a very short time they developed into the fledged condition and forthwith set about their devastating work in right good earnest as the following report from the *Melbourne Age* of March 23, 1891 will show:

AN AUSTRALIAN CONFERENCE RE-LOCUST PLAGUES.

Said report refers to an extraordinary conference of leading vignerons orchardists, agriculturists and shire councillors held in the Royal Victorian Society's rooms, Melbourne, on the 20th of said month, at which I was present. The object of the conference being to, if possible, devise legislative or other means to save the colony from continuous visitations of destructive locust blizzards:—" The chairman opened the meeting by reading a " circular suggesting that the conference should be held, signed by Mr. Samuel Trethowan of Nathalia, from which circular the following is an extract. " In view of the deplorable destruction caused throughout the colony by the locust plague, I feel impelled to seek your valuable assistance in devising means to prevent a repetition of such disastrous invasions," (as during Dec. of the previous year). " I have been a settler in the colony for thirty-five years, and never before saw such myriads of locusts as have just passed over the land; devouring every green leaf and shoot. I believe the time to be opportune for the taking of energetic and concerted action to destroy these voracious insects before they turn our homes and holdings into permanent scenes of bare and naked desolation. I have, myself suffered to the extent of losing every leaf, shoot and bud on an orchard and vineyard of one hundred acres in extent, besides the loss of grass and substance for my stock on several thousand acres of land, and also the loss of a considerable portion of my crop of grain as well. By the visitation of this scourge a crushing blow has been dealt to the prospects of intense culture in this fertile but insect infested region, and the Government may well be asked to stay any further expenditure on vast irregation schemes until some means are found to repress this terrible pest, for the result of years of patient labor of an industrious settler may be ruthlessly destroyed in a few hours by these destructive insects. I am sure I am within the mark when I estimate the loss to the colony to be several hundreds of thousands of pounds, for, apart from the destruction of gardens, orchards and vineyards, graziers have been compelled to sell their stock at a sacrifice, in consequence of the destruction of everything that could be eaten in the paddocks by these ravenous pests. Practically the whole of the northern part of Victoria is now denuded of every succulent blade or shoot and the choicer and more careful the cultivation the greater has been the havoc made.

" Something must be done or the land will become an uninhabitable wilderness, for the locusts are now breeding, and before the summer is over they will be upon us again in even more overwhelming numbers than those just passed over. The rabbits have been and are a terrible scourge, yet their destructiveness pales into insignificance when compared to the ravages of the *vast armies* of winged and wingless locusts.

" The scourge is an Australian and the question of legislation a federal one demanding simultaneous action by the governments of the several colonies in order to be effectual." " The chairman said he agreed with

Mr. Trethowan....We are threatened with a far greater invasion by the locusts, an invasion which, if not checked by some means, would eventually devastate the whole colony." The conference unanimously endorsed the chairman's opinion and appointed a large representative committee to confer with the Hon. the minister of agriculture on the subject.

The three leading agricultural colonies within one month suffered from the said locust visitation to the extent of about one million eight hundred and fifty thousand pounds (*i. e.* $9,250,000) as follows—New South Wales £450,000 ($2,250,000), Victoria £600,000 ($3,000,000) and South Australia £800,000 ($4,000,000). Queensland also suffered much from the invasion. Since then the Chaffey Brothers' magnificent Mildura irrigation colony in Victoria was invaded by dense clouds of locusts and for nearly a whole month seemed as if they meant to permanently locate there and in the immediate neighborhood, in order to dispute the Chaffey Brothers' right to have transformed that once arid barren region into a beautiful fertile garden.

A FALSE REPORT.

About the time the above locust trouble overspread the Mildura district a very silly theoretical suggestion was made by the chief Entomologist in Victoria to the local government at Albury, N. S. W., namely, to erect screens over trenches in order to stop the plague, coupling the suggestion with an absurd estimate of the cost per mile of such, which, when enquired into was found to be utterly useless if adopted, besides being greatly underestimated as to cost. On the scheme being ridiculed by an expert through the local press, the Victorian secretary for agriculture acknowledged his entomologists error and withdrew the proposition. After the dense clouds of locusts had devoted the greater part of a month gorging themselves with the very choice products in the Mildura estate, they passed on towards the adjacent colony of South Australia, leaving a few hundreds scattered about, dead and dying, from possibly, the eating of tomato plants. About sixty of the dead were found to be fly-blown, which fact was carefully figured up by a humorous newspaper man, who wrote a sensational paragraph concerning "a wonderful locust destroying fly," etc., which paragraph has evidently been seriously considered by the State of California Horticultural Department, as the following from the fertile pen of the State Entomologist published in the *S. F. Examiner* of February 2nd last, denotes:—" An effort has been made among others, to introduce a parasite of the grasshopper which is found in Australia, where it has accomplished very excellent work, this is a fly a species of Tachina, which feeds as voraciously on the grasshopper as the latter does on vegetation, and has contributed very much toward keeping down that pest in Australia (!) The egg is deposited by the female in the body of the grasshopper and hatches there. The young grub lives upon the adipose tissue of the victim and avoids the vital part until it is matured, sometimes several of these grubs may be found in the grasshopper. When full grown the grubs eat their way out of their victim, usually at the side where the abdomen and the matathorax meet. As soon as the grub escapes, the grasshopper, which has been growing weaker as his parasite has grown in side dies. The grub then buries itself in the earth and undergoes its transformation, immerging a perfect insect. Examination proved *in one case* that sixty or seventy per cent. of the grasshoppers in one part of Australia were parasited by the insects."—(Not one of those wonderful flies have been as yet captured).

During my observations anent the early invasion of said tiny grass-hoppers, I noticed that their onward progress towards the settled districts was invariably impeded by the existence of any intervening clumps of eucalypti saplings or trees, as also scattered eucalypti leaves causing the whole mass to make wide detours away from such. I was thereby led to more carefully examine the reasons why this was thus, and ultimately discovered—subsequent to having experimently protected my vine-yard and orchard grounds with Eucalypti plants, that the marvelous antiseptic exudations from the foliage absolutely defy the intrusion of locusts or of other insect plagues including the much dreaded phylloxera vastatrix, if vines are reasonably well planted and cared for, it being im-possible for locusts or grasshoppers however numerous and powerful they may be in and around the immediate neighborhood to approach within thirty or forty feet at least, of eucalypti protected grounds and live. This fact I have satisfactorily, though privately, demonstrated for over a decade in Victoria, Australia, and am fully prepared to do so in the United States or elsewhere.

In my researches into the origin of insect plagues, I have discovered that they are as already stated "nature's scavengers" evolved and sus-tained from the putrid atmosphere which our reckless destruction of the earth's "forest lungs" has created, and that live eucalypti foliage is un-questionably the finest and most perfect atmosphere purifying agent on this blight cursed earth. Just think of it, Eucalyptus glolulus leaves are sold by wholesale druggists in Victoria, Australia at one shilling and sixpence per pound, *i. e.* thirty-six cents, for medicinal purposes, and in California they are, as a rule, ruthlessly chopped down and trodden under foot!

U. S. CONSUL E. L. BAKER'S REPORT.

In the United States Consular Reports for November and December, 1882, appears an interesting account from Consul E. L. Baker on the beneficent properties of eucalypti trees in Buenos Ayres, dated August 23rd, 1882, from which I have pleasure in quoting as follows :

THE EUCALYPTI TREE.—"Thinking the matter might be of some interest to the people of the United States, I enclose a report setting forth the success which has attended the introduction of the *Eucalyptus Globulus* in the Argentine Republics and somewhat explaining the method of its cultivation here. I have several times in my annual reports referred to the successful introduction of the *Eucalyptus Globulus* (the blue gum tree) of Australia into the Argentine Republic, and spok-en of the rapidity some portions of the pampas heretofore destitute of timber are now being dotted with plantations of these magnificent trees ; and from the ease with which they can be grown, and the marvellous rapidity of their growth. I suggested the possibility of their cultivation in the milder parts of the United States. I observe that in a more recent report to the Department (No. 8, page 890) accompanied by an article from a French paper, Consul Wilson of Brussels shows the success which has attended their introduction into France, and also sug-gests the acclimation on the treeless regions of our southern territories. From what I have observed during my stay in this country, I am more and more con-vinced that the eucalyptus is a most desirable tree with which to timber our south-western plains and renew our rapidly decreasing forests; and I believe that a proper trial on a large scale in those portions of our country where the winters are not too severe, would speedily render it such a favorite that its cultivation, not only for ornament, but for timber would become general.

"VARIOUS SPECIES OF EUCALYPTUS.—To avoid disappointment, however, in at-tempting plantations in the United States, it should be borne in mind that there are many species of this tree ; and that not all or even many of them would bear our climate except, perhaps, that of our extreme Southern States, as they require a tropical or sub-tropical temperature. Others of them, however, are quite hardy and *are capable of bearing very severe cold*. With a view of testing their cold bearing qualities Dr. Ernest Aberg, a member of the Academy of Medical Science, and a German scientist who has resided many years in Buenos Ayres, has lately devoted a large share of attention to the growing of these trees. He has now at his county-seat, near the town of Ramos Mejia, on the Buenos Ayres and Western Railroads, upwards of sixty varieties in more or less successful cultivation.

"In a recent publication on the subject he says, that while the most of them are too delicate to be grown in the province of Buenos Ayres, except as ornamental trees, requiring special care, *a few are not affected by even severe cold* and can bear the changes of the most variable climates. Amongst these he specially recommends the following : *Eucalyptus alpine*, which grows on the highlands and mountains of Victoria at an altitude of 4000 feet above the level of the sea. *Eucalyptus amygdalina*, a very large tree, measuring generally upwards of 80 feet in height by 2 feet in diameter while it is not rare in Australia to see specimens 200 feet high and 40 feet in diameter at 4 feet from the ground. The tree seems to do well in Buenos Ayres, growing very rapidly, and producing a magnificent hardwood with beautiful veins running through it. Its vulgar name among the colonists is the "narrow-leaved peppermint tree." *Eucalyptus coriaca*, a tree which endures the intensest cold, forming in New South Wales and Victoria extensive forests at an altitude of 5000 feet above the sea level. Its common name is the " mountain white gum tree." Its leaves are large and lustrous.
. . . . *Eucalyptus globulus*, this tree is not quite so hardy as some of those mentioned above, especially when it is young, but its wonderful qualities have made it naturally a great favorite here, and it is now to be seen growing sometimes on large plantations in all this part of the province of Buenos Ayres, its very rapid growth in two or three years producing the same effect which other trees would hardly produce in fifteen years. This is the variety of *Eucalyptus* of all others most recommended for acclimation in the United States. The *Eucalyptus globulus* grows with a rapidity which is surprising . . . . as an example of their increase it may be stated that in Hyères seeds planted in 1857 had in 1865 reached the height of 58 feet. In Toulon the plant grows to 24 feet in two years. In 1863, there were trees in Algiers of three years growth which had attained a height of 30 to 35 feet, and generally in that country they grow at the rate of about ten feet each season.

"MEDICINAL VIRTUES OF THE TREE.—I have said this much of the *Eucalyptus* as a forest or shade tree which has already become a great favorite here, and which, I believe, would be found to give equally satisfactory results in certain parts of the United States. I may add that it is considered here *to be a very healthful tree*. The pungency of its leaves is such that *it is never molested by insects, and I believe it is the only tree grown here which the locusts* (the great pest of the Argentine Republic) *will not attack*. It has the reputation of being *an excellent destroyer* or absorbent of malaria. It is stated that in Australia there are no marsh fevers where large forests of the *Eucalyptus exist*" (perfectly true) " and I hear they have been planted in the Pontine marshes near the city of Rome with excellent effects. In regard to its medical virtues, Professor Bently published a pamphlet in London in 1874 in which he fully confirmed all that has been claimed for the tree, calling it "THE FEVER DESTROYING TREE," and citing many instances to prove it. Here in Buenos Ayres they bruise the leaves and bind them to the forehead for nervous headache, and I am told that the leaves are *a special abhorrence to such insects as prey upon fruits and fruit trees against whose visitations they furnish protection by being scattered thickly on the ground underneath.*"

The above extract which I have copied since my arrival in San Francisco, speaks for itself and, in my opinion, should be read from every church pulpit and school desk in America until the instructive lesson it conveys be deeply burned into every one's mind throughout the land. Consul Baker's admirable report had evidently been overlooked and became in consequence "a dead letter" to those most concerned, viz. vineyardists, orchardists and agriculturists, etc.   Shortly after my arrival in San Francisco I learned from D. Henshaw, Ward Esq., of 308 California Street, general manager of the Natoma (1500 acre) vineyard, that within two months in 1891, the company he represents suffered to the extent of $65,000 from a grasshopper invasion; we shall therefore be correct in surmising that all other vineyards and orchards within the infested area, suffered in like manner, thereby possibly aggregating a loss amounting to sundry million dollars in this State alone which could have been saved had said report been made known and acted on.

The testimony of Australia's premier authority—Baron Ferdinand von Mueller, K. C. B. M. D., F. R. S.—on the nature and properties of Eucalypti, as expressed in the following extract from a letter dated December 7th, 1883, is now of special interest :

" The European and American mails, dear Mr. McLean, kept me incessantly engaged till Tuesday night, and since then I have been harassed with official work; hence I can only now at a late night hour attend to your inquiries. . . . What the bottoms of temporarily dry lakes would generate are Bacteria in billions sometimes as the carriers of epidemic diseases. . . If therefore your wise suggestion could be adopted to let the water of the Avoca into the Lake Tyrrell (an immense waterless basin in an arid district) not only would ample drinking water be gained but the air be cooled far around and the bacteria be drowned.

" As regards planting for sanitary purposes, *nothing can be more valuable than Eucalyptus*, their odorous foliage originate ozone and peroxyde of hydrogen as most powerful destroyers of miasmatta.   The little book transmitted herewith (a treatise on *Eucalypti*) for your kind acceptance may interest you as bearing on questions in which you have shown such a deep interest."

" With regardful remembrances, yours

(signed)   FERD. VON MUELLER."

FURTHER TESTIMONY

From the *Bendigo Advertiser* (Australia) of October 23rd, 1890 :—

GUM TREES AND DRAINAGE.—" The value of eucalyptus trees for draining swampy land is illustrated by the following paragraph from the *St. James Gazette:* " For years past the Trefontane Convent at Rome had become positively uninhabitable owing to the malaria which attacked—in many instances with fatal results—its inmates.   Senator Torelli presented a bill in Parliament proposing that the estate annexed to the convent should be planted with eucalyptus as an experiment against malaria,   The bill was passed, and the Trappist monks planted thousands of eucalyptus plants of all species on the estate.   But still the malaria ravaged, and several monks suffered severely.   But it was remarked that it was only the monks who had their cells looking on the central cloister who fell victims to the malaria.   This suggested the idea of planting four eucalyptus trees at the four corners of the cloister.   The plants, sheltered from the winds, soon grew to a great height.   The immediate result was the complete draining of the soil in the cloister, and the disappearance of malaria fever from the convent."

That eucalypti is unquestionably the most powerful antiseptic tree on earth cannot now for a moment be seriously doubted, and when we couple its many additional virtues including its rapid growth, durability and beauty of texture for furniture and general building purposes—(as will be amply shown by the New South Wales Government at the Chicago World's Fair)—it is, in my opinion, justly destined to play a leading part in forest restoration all over this planet.   The almost limitless medicinal properties which permeate its evergreen " odorous foliage" are being rapidly demonstrated to the discomfiture of all disease breeding germs. May we not therefore hope that we have found in the leaves of this beneficent tree a real panacea "for the healing of the nations," which should be liberally planted under the very best local supervision through and around every populous centre where sound health is desirable? and in order to encourage their most pleasing growth, an annual prize might be awarded for the best kept plot of trees entrusted to the care of adjacent residents, on a public "arbor day." Such an arrangement would quickly transform all malaria breeding cities of the Chicago type into healthful retreats. If the streets were ornamented with *eucalyptus fosifolia* and *eucalyptus globulous* (or a more hardy sort such as *eucalyptus amagdalina*) planted alternately, a most pleasing result would follow, as the *fosifolia* evolves a profusion of beautiful purple flowers.

The whirlgig of excitements connected with huge business concerns throughout the United States and elsewhere, seems to have deadened every other consideration, and therefore doubtless the scribbler of this treatise will be considered a silly alarmist. There is, however, one thing certain, namely, that ere long the great ones of the earth will be compelled to earnestly turn their attention to the terrible results now hourly increasing from outraged nature which seriously threaten to depopulate the apparently fairest parts of this planet. As shown by Professor Ellis—we, for over five thousand seven hundred years, according to the Mosaic record of time—had been ignorantly slaughtering the earth's forest lungs, before Dr. Priestly, in his scientific researches discovered in them the hidden atmospheric regulating treasures, yet notwithstanding his and others admonitions; since then we have continued the devilish work in a more vicious manner. Australia now suffers immensely from consequent droughts, floods, fires and various insect-plagues, causing universal depression amongst pastoralists, miners, agriculturists and artisans, and hence the existing wholesale insolvency through the colonies extending to the leading banking corporations—the latest being one of the most gigantic of those institutions, about which the following dispatch from the *S. F. Examiner* of April 13th, 1893, speaks:—

A BIG BANK FAILURE.

LONDON, April 12.—The English, Scottish and Australian Chartered Bank has failed with liabilities amounting, it is said, to £8,000,000 or $40,000,000. No estimate of the assets has yet been made, but they are supposed to be large. The bank was incorporated by royal charter in 1852, and claimed to have a paid up capital of £900,000, and a reserve fund of £310,000. The suspended bank has its main branches at Sydney, Adelaide, Brisbane and Melbourne and at various lesser points in the colonies of New South Wales, Victoria and South Australia. It transacted a banking and exchange business between Great Britain and the Australian colonies, and had large deposits. The failure has added to the anxiety and consternation which recent failures of financial institutions, with Australian connections, have caused. The only reason given for the failure is that there has been, for several weeks, a steadily increasing withdrawals of deposits.

ADDITIONAL EVIDENCE OF RUINED DEFORESTED SOIL.

By the *S. F. Chronicle* of April 16th, 1893, we learn that the Pacific Mail Steamship *China* arrived on the preceding day, bringing Hong Kong news to March 18th and Yokohama advices to March 25th, and that the news consists mainly of disasters by land and sea; how that the people are selling their children to get money for wheat, owing to the existence of a widespread famine in Mongolia and in Shansi, declaring that the outlook is even worse than in the memorable year that followed the big overflow of the Yellow river—"For 150 miles not so much as a single grain of wheat has been reaped by the inhabitants, so that after they had consumed their reserves nothing more could be done than to lie down and die." Here we have unmistakable evidence of ruined soil in an almost treeless portion of a vast Empire, soil that has been cropped for centuries to feed teeming multitudes minus any recuperating rest, and, as stated by M. Marchand in his work I have already quoted :—"With the ruin of the soil begins the ruin of the people. The more unhappy they are, the more selfish do they become." And this is how Marchand's affirmation is borne out in this case. The *Chronicle's* correspondent makes reference to the miserable appearance of the famine cursed people, as also to the heartless treatment they are being subjected to from their more wealthy brethren in the following manner :—

"They reminded one on looking at them like so many skeletons, with faces as sharp and pointed as the eagle. The rich in the famine districts stayed at home, of course, but they had to economize their scanty stocks of oil and wheat, while those either too old or poor to leave the country were fortunate if they could find herbs and roots of trees and plants to satisfy the cravings of starvation. He must have had a heart of iron not to have wept tears whenever his eyes rested on the hundreds of bodies lying on the roads and by-paths, either in a dying condition or stiff and lifeless. And the bitter cold! Which of these two cruel enemies took away the most—cold or hunger? The crafty grain dealers of Shansi made enormous profits out of the poor sufferers, some of whom, having no money to buy the precious cereals, sold or gave in exchange their children in barter. For instance, a child of 6 was worth many hundred cash, and marriageable girls were bartered in exchange for a camel's load of wheat or 400 catties.

"A correspondent from North Shansi says that Dr. Stuart reports that near Ningwu Hsien he met seventeen loads of young women and girls on the way toward the south to be sold. Each load had an average of twenty persons, all from one district. The people of the Kueihua Hsien villages say that out of every three persons two will die before the end of the second month of next year. This year the oat and wheat crops were practically a failure, and the millet crop was not more than one-half as large as usual."

The *Chronicle* also alludes to fearfully destructive fires and snow storms in Japan and in Canton causing great loss of life. 200 persons perished in Canton—crushed to death by the immense weight of snow falling on their dwellings and in the streets. "Snow was a thing unheard of in Canton," adds the writer, etc. Surely "grievous times" are now upon us !

A DEFORESTATION.—LESSON FROM RUSSIA.

"The first article in the *Edinburgh Review* for January last is entitled "The Penury of Russia." A more dreary and unrelieved picture of blank desolation has hardly ever been printed." (*Review of Reviews*, March, '93).

FORESTS AND RAINFALL.—Without entering into details, here is one startling statement made by the reviewer. He says that owing to the destruction of the forests the rivers are drying up, and the eastern part of the country is literally being sanded up : "The ruthless forest destruction which has been going on for a long time has had a serious effect in reducing the average rainfall. The belts of wood attracted and held the moisture, which was slowly distributed for the benefit of agriculture ; now, in vast regions, as, for instance, on the black soil, there is hardly a tree to be seen, and the consequence is that the underground rivulets which nourished the soil have disappeared. The forests also broke the force of the fierce east desert winds. Now these winds, piercingly cold in winter and scorchingly hot in summer, burst with full fury on the great plains. In summer their blasts are capable of withering the corn in a few days and with them come sand storms, which turn fertile land into permanent deserts. The unfortunate experience of Central Asia, which once was a garden of fertility and now is a desert peopled by nomads only, are repeating themselves.

DRIFT SAND FROM THE DESERT.—"In the province of Astrachan an area of 800 square miles is covered by drift sand ; in that of Strawropol whole villages have disappeared, and in 1885 soldiers had to be summoned to clear the sand from the houses. In the province of Tauris the sand now covers 150,000 des-jaetines (1.00925 hect); the same disastrous effects took place in the north, where, after the destruction of the forests in the provinces of Samara, Woronesh and Tchernigow, hundreds of sand hills arose, which gradually covered the fertile land. A further consequence is that the rivers become shallower. In winter there is nothing to hold the snow, which is blown together into large heaps; these with the thaw dissolve into temporary torrents, washing away acres of tillage, and carrying off all moisture before it has had time to soak into the soil.

THE DRYING UP OF THE RIVERS.—"The river beds cannot contain all this water, and inundations occur ; but when it has swept down there is no further supply. The Woronesh, on which Peter the Great built his first ships, is now a mere rivulet ; the Worskla, which fifteen years ago was a beautiful river, surrounded by woods and pastures, has absolutely disappeared ; the Oka has become so shallow that barges coming from Nishegorod were stranded upon its sands. At Dorogobush the Dnjepr can be crossed by carriages; on the Dnjepr the navigation had to be stopped, as its depth was reduced to 2 or 3 feet; and even on the Volga steam navigation is interrupted in many parts, the river not being able to carry away the sandbanks; it is calculated that the volume of its water has decreased by 24,000,000 cubic meters. It is evident that even the most costly works for opening the channels will be of little avail; the cause lies in the devastation of the forests; the law by which the government interdicted the ruthless fall of timber has come too late, and replanting is slow work, although *it is the only remedy against the evil.*"

The following reliable outline—with copies of photos—of an American mischief creating Sahara, appeared in the San Francisco *Chronicle* of April 9, 1893. The writer, a Mr. Frederick I. Mousen, recently returned to San Francisco after a three months exploration tour through " Mojave Desert and Death Valley," about which he now writes:—

" Death Valley is known as the region of lowest depression in the world, besides claiming the flattering appellation of being the hottest place on earth. It is 430 feet below the level of the sea. The valley is seventy-five miles long and from eight to fifteen miles wide." (A splendid breeding field for insect plagues and their broadcast distributing whirlwinds.) On the east the valley is bounded by the Funeral mountains, which attain an elevation of from 6000 to 8000 feet, and on west it is inclosed by the Panamint rage, which reaches a height of from 8000 to 10,000 feet. "The valley is an independent drainage basin, and the eastern part is filled with a wash of rock and gravel, the result of cloudbursts. Immense fields of borax and soda cover a large section of country in the eastern part of the valley, and salt marshes of the almost pure chloride extend over a vast area of land. From a spring in Furnace creek wash, the entrance to this arid country, the Pacific Coast Borax Company cultivates about thirty acres of land in alfalfa, the only evidence of civilization in the entire district. Were it not for this ranch it would be well nigh impossible to make the trip across the valley, as by no other means could feed for the horses be obtained. The nearest accessible point to Death Valley is Daggett, a small station on the Atlantic and Pacific Railway. It is 105 miles distance from this point to the valley, and requires a journey of seven days to cover the ground. On the road there are but three springs, two of which are sixty miles apart. Travelers are, therefore, compelled to carry water for themselves and beasts, and when it is added that one has an inordinate thirst on the desert the burden can be considered no light one. A man will drink three gallons of water a day and the animals twice as much as customary. But little good water is found, as most of the water holes or springs are charged with alum, arsenic or borax. On the following day after my arrival at Daggett, I left, equipped with an outfit consisting of buckboards, mules and a guide. We crossed what is known as the link of the Mojave river and journeyed for five days over a region destitute of vegetation or animal life, with the exception of a growth here and there of dwarfed cacti. The oppressiveness of this desolation and extreme solitude must be experienced to be understood. Nameless graves of poor unfortunates who attempted to cross the desert during its heated term are the only break in this dreary monotony, and every year new mounds of earth, marked only with a stick or a stone, show the spot where some adventurous prospector perished from thirst and the excessive heat and was buried by strangers.

"Desert travel during the summer months is attended with extreme danger, and can only be accomplished by traveling at night and camping during the heat of the day at some water hole or small oasis. If an accident had occured to our wagon or mules on the desert it would have been a very serious proposition as we were miles from any human habitation and it would have been impossible to have secured other animals or repair a break. On the fifth day out we reached Armargosa, the abandoned borax works of the Pacific Coast Borax Company, and here rested for a few days. It is sixty-two miles from this place to Death Valley and we had to carry water for this entire distance. We occupied two days in this last stretch and we travelled the entire route over a bed of rock and gravel, accumulated by the action of cloudbursts.

"The entrance to the valley is through a canyon called Furnace-creek wash. We arrived here just before dusk and at an elevation of 5000 feet we obtained the first view of this historical valley. Far toward the west the Paniment mountains stood, forming the wall of that side of the valley. Just behind these western ridges was sinking the ruddy sun, bathing tnis desolate production of God's hand (?) in a purple-tinted light. Moment by moment the shadows crept over this scene of desolation. No sound from the twittering lark or the wild canary forewarned us of the approaching night. All was as still as the midnight hour. We prepared our camp, exchanging hardly a word and glad for the moment when we would become oblivious to these ghostly surroundirgs. The next morning we arrived at the entrance or level of Death Valley and from this point saw for the first time a sand storm. Incidentally I may remark that these storms during the summer months are very frequent. When we saw it the northern part of the valley was obscured in a brownish-colored cloud, which gradually thinned out until it crossed the entire valley. Never at any moment did it entirely obscure the sun, which looked like a mere ghost of itself. The cloud moved with great velocity in our direction and soon advance runners or hot puffs of wind reached us. On arriving at the end of Furnace-creek wash we saw the sand storm in the northern part of the valley. This was shortly after sunrise. The entire horizon and sky was obscured by the sand and gravel, which were buoyed in an atmosphere oppressively hot and stifling.

"Soon we were enveloped by a dense cloud of sand, and occasionally as a stronger puff of wind came gravel and even small rocks were hurled into our faces. We covered our heads with blankets, and the mules instinctively turned their tails to the wind. With all these precautions we did not escape, for my guide's face and my own became badly bruised and lacerated. In our exposed position we experienced all the fury of this desert simoon, and as the winds traveled across the alkali sink they increased in heat to such an extent that breathing became a matter of difficulty. This storm lasted two days, and in all this time it was impossible to move from our position. We had to camp here for that period, being unable to light any fires or prepare meats. Most of this time we were covered with blankets and literally starved. I never want to undergo that ordeal again. This was in the fall. The puffs of wind were so intensely hot and suffocating that they can be likened to blasts from a furnace, and seemed to draw the very breath from our bodies. The storm piled the sand around our wagon and covered everything. Occasionally looking down into the valley, we could see a large sand augur or spout waltzing hither and thither over the country, carrying the sand and alkali dust *high into the air.* Though this storm was exceedingly severe to a novice like myself, it is incomparable to the ones that sweep over this country in middle summer. In July or August, with the thermometer registering from 130 to 137 degrees in the shade, it would be impossible for any living creature to exist in it even for an hour."

If "Death valley" was flooded from the sea or elsewhere, and surrounding mountains properly clothed with our evergreen eucalypti, one very great factor for good to thousands of struggling settlers and others far and wide would most assuredly ere long follow, as the billions of disease breeding germs now annually evolved in said valley and widely scattered broadcast over land and sea by whirlwinds, would be drowned and the surrounding air delightfully cooled, besides considerably lessen the existing destructive storm creating cause. Some such extreme measures must shortly be taken to transform American and other deserts from their present dangerous condition, if we really wish to abate the now hourly increasing cyclonic and torrental devastations all over the earth, for notwithstanding our long practical denials that we are our brothers' keepers, the immutable laws of nature—which are the laws of God—teach a very different tale. The question has often been and is now asked : " Why is it that Europe should have been so severely afflicted by the cholera and other disease scourges from Asia ?" And all sorts of surmises have been ventured on regarding the transmitting agency of said scourge—just as we now ponder anent the impending cholera epidemic from Europe to the United States and Australia—quite oblivious to the fact that we commenced our forest-lung destroying work in Asia, and that the consequent " oriental afreets" or whirlwind sand-pillars of the desert, have been and now are the real transmitting agents which charge the " ariel reservoirs" with myriads of disease germs "seeking whomsoere they may devour," often flooding ships in mid-ocean.

IMMENSE VALUE OF THE AMERICAN GRAPE GROWING INDUSTRY.

(*From Harper's Weekly, July 18th*, 1891.)

A special investigation shows that in the several grape growing districts in the United States 401,261 acres were set apart for the industry, 307,575 acres in bearing producing 573,139 tons of grapes, and 240,450 tons were used in producing wine, making 24,306,905 gallons, 41,165 tons for raisins, making 1,372,195 boxes of twenty pounds each, and 23,252 tons for dried grapes and purposes other than table fruit. The total value of plant used in the industry, 1889, is given as $155,661, 150; at the time of taking the returns 200,780 persons were employed.

OTHER FRUITS.

(*From the San Francisco Examiner of March 24th*, 1892.)

" As the result of a careful compilation of information from the most reliable sources, the growers themselves, the Census office has issued some interesting figures upon the production of oranges, lemons, figs, almonds, cocoanuts and other semi-tropical fruits and nuts in the United States. Numbers of acres of bearing and non-bearing trees and plants, 271,428.10; number of bearing trees and plants, 28,101,036; number of non-bearing trees and plants, 14,205,323; value of product for year 1889, $14,116,226 59; estimated number of acres suitable for planting tropical fruits and nuts, 24,710,879.

The comparison between California and Florida in relation to fruit trees is interesting. In this State there are 78,616.47 acres devoted to bearing and non-bearing trees. Florida has 168,754 63. California has 2,652,021 bearing trees or plants, while Florida boasts of 25,317,536. California has 4,247,789 non-bearing trees and Florida has 9 200,764. California has 38,367 acres devoted to oranges, while Florida has 144,769. This State has 1,153,881 orange trees bearing, against Florida's 2,725,272; non-bearing orange trees in California, 2,223,710; Florida has 7,408,543.

LOCUST PLAGUES IN ALGERIA.

It is said that as a result of early forest destruction Mahomet was inspired to pen the following "fable" :—

" We are the army of the great God," quoth the locusts. " We produce ninety-nine eggs each. If the hundred were completed we should consume the whole earth and all that is in it."

And it is worthy of remark that the earliest recorded invasion in Europe of these insects, A. D. 591, was from Africa, the northern shores of which has been rendered bare and sterile by centuries of forest destruction. The *Scientific American* of July 25th, 1891 furnished the following interesting particulars regarding recent terrible results in and around Algeria:

" During the past three or four years the French Government has been making strenuous exertions to beat down the armies of locusts coming from the South on the fertile lands of Algeria, and during the present year they are also having a similar fight with these pests on the northern borders of Lunis. The cheap Arab labor obtainable for this purpose has made it possible to employ in the work a veritable army of men, the Government ordering the tribes to form encampments along the line on which it is proposed to fight the oncoming army of locusts and in this way the crops have been protected from the ravages of this plague, *although no permanent relief has been obtained.*"

And the following from " *Land* and *Water*" of August 8th, 1891, still further testifies to the fearful scourge in that country :

" The attention of the Board of Agriculture has been drawn by the Foreign Office to a report from her Majesty's acting charge d' affairs at Tangier, (Algeria) respecting the severe effects of the visitation of locusts on the crops in Morocco. From this report it appears that the damage done by these insects in the chief grain growing districts has been very considerable, the latter crops having suffered to a very large extent. In the Tangier district and in the northern province the crops have been but little affected owing to the late arrival of the locusts, but the injury became more extensive towards the south. Thus at Rabat and Daralbaida half the wheat crop has been destroyed. At Mazigan the maize and pea crops are complete failures, while at Mogador there is a general scarcity of grain. Olive and almond trees have suffered extensively in most districts, oil being only sufficient for local consumption. From all parts of the country fruit and vegetables are reported to be *entirely destroyed.* In addition to the loss of the crops of grain, fruit and vegetables, a serious feature is indicated in the want of pasturage, which has produced a mortality amongst the cattle, and a great fall in their price, owing to the anxiety of the people to sell animals they are unable to feed."

Whether or not the famous Mahomet " fable" anent the destructive power of locusts has any foundation in fact, the following from the *New York Herald* of May 18th, 1891, unmistakably denotes their apparently revengeful disposition:—

KILLED BY LOCUSTS.

" The horrible death that befell a French savant, M. Kunckel Herculass, the president of the Ethnological Society, who was employed on the Government mission investigating the locust plague in this province, has met with a horrible death. While examining a deposit of locust eggs at the village of Sidiral, (Algeria) he was overcome with fatigue and the heat and fell asleep on the ground. While sleeping he was attacked by a swarm of locusts. On awakening he struggled desperately to escape from the living flood. He set fire to the insect laden bushes* near him, but all his efforts proved ineffectual, and when finally the locusts left the spot his corpse was found. His hair, beard and neck had been entirely devoured. M. Herculass was a member of the French Academy and author of several valuable works on insects.'—(*smoke stimulates locusts).

"On the 13th of May last," wrote a lady contributor to the *Contemporary Review* for June, 1891:—"I was traveling with my husband through Eastern Algeria. At six o'clock on a lovely summer's morning and there, before us in the transparent air, looking like a summer snow storm, we saw approaching a dancing cloud of winged particles. It was the advance guard of the dreaded locust army . . . The whole of this wide expanse—including the three departments of Oran, Algiers and Constantine, which composes the colony stretching from Morocco on the west to Tunisia on the east, the city of Algiers standing about half way between the two boundaries, and the whole coast line being about a thousand kilometres in length—is threatened with ruin, ruin compared to which the ravages of the phylloxera are mild. The last news that we have from the Western Province was that around Flemeen, on the frontier, flights of locusts were alighting unintermittently, and that a caravan just arrived there from Morocco, had travelled for thirty days in the midst of locusts, *the country being entirely devastated.*"

SCRIPTURE WARNINGS.

Surely the warnings given through the prophets Moses and Joel were not meaningless dreams? Warnings of results that were sure to follow in the wake of wilful disobedience to nature's teachings—results foreseen by Him whose servants said prophets were and who caused them to write as follows:—"Thou shalt carry much seed out into the field, and shalt gather little in; for the locust *shall* consume it. Thou shalt plant vineyards and dress them, but thou shalt neither drink of the wine, nor gather the grapes, for the worms shall eat them. Thou shalt have olive trees throughout all thy borders, but thou shalt not annoint thyself with the oil, for thine olive shall cast its fruit. All thy trees and the fruit of thy ground shall the locusts possess."—DEUT. 28. And after centuries of forest destruction Joel wrote:—"Hear this ye old men, and give ear all ye inhabitants of the land. Hath this been in your days, or in the days of your fathers? Tell ye your children of it, and let their children tell their children, and their children another generation. That which the canker worm hath left hath the locusts eaten; and that which the locusts hath left hath the canker worm eaten; and that which the canker worm hath left hath the caterpillar eaten, for a nation (of insect plagues) is come upon our land strong and without number; his (locusts) teeth are the teeth of the lion, and he hath the jaw-teeth of a great lion. He hath laid my vine waste, and barked my fig tree, he hath made it clean bare, the branches thereof are made white. . . . Be ashamed all ye husbandmen, howl, O ye vine dressers, for the wheat and for the barley, for the harvest of the field is perished, the vine is withered and the fig tree languisheth; the pomegranite tree, the palm tree also, and the apple tree, even all the trees of the field are withered; for joy (the Father's guiding love) is withered away from the sons of men. The seeds rot under their clods, the grasses lay desolate, the barns are broken down, for the corn is withered. How the beasts do groan! the herds of cattle are perplexed because they have no pasture!"

NEWMAN'S CALLISTA.

"The plague of locusts, one of the most awful visitations to which the countries included in the Roman Empire were exposed, extended from the Atlantic to Ethiopia, from Arabia to India, and from the Nile and Red Sea to Greece and the north of Asia Minor. Instances are recorded in history of clouds of the devastating insects crossing the Black Sea to Poland, and the Mediterranean to Lombardy. It is as numerous in its species as it is wide in its range of territory. Brood follows brood, with a sort of family likeness, yet with distinct attributes, as we read in the prophets of the Old Testament, from whom Bochart tells us it is possible to enumerate as many as ten kinds. Even one flight comprises myriads upon myriads, passing imagination, to which the drops of rain, or the sands of the sea are the only fit companions; and hence it is almost a proverbial mode of expression in the East (as may be illustrated by the Holy Scriptures), by way of describing a vast invading army, to liken it to the locusts. So dense are they, when upon the wing, that it is no exaggeration to say that they hide the sun, from which circumstances, indeed, their name in Arabic is derived. And so ubiquitous are they when they have alighted on the earth, that they simply cover or clothe its surface.

" This last characteristic is stated in the sacred account of the plagues of Egypt, where their faculty of devastation is also mentioned. The corrupting fly and the bruising and prostrating hail preceded them in the series of visitations, but they came to do the work of ruin thoroughly. For not only the crops and fruits, but the small twigs and the bark of the trees are the victims of their curious and energetic rapacity. Nor do they execute their task in a slovenly way, that, as they have succeeded other plagues, so they may have successors themselves (such as the canker worm, commonly known as the " measuring worm" now so destructive in California.)

"They take pains to spoil what they leave. Like the harpies, they smear everything that they touch with a miserable slime, which has the effect of a virus in corroding, or as some say, in scorching and burning. And then, perhaps, as if all this were too little, when they can do nothing else, they die, as if out of sheer malevolence to mankind, for the poisonous elements of their nature are then let loose and dispersed abroad, and create a pestilence; by which they manage to destroy many more by their death than in their life. (After locust blizzards in Australia great ridges of the dead are generally deposited on the sea shores of South Australia, of inland lakes, and of Hobson's Bay in Victoria, creating a fearful stench).

"Such are the locusts—whose existence the ancient " heretics" brought forward as their primary proof that there was an evil creator, and of whom an Arabian writer shows his national horror, when he says that they have the head of a horse, the eyes of an elephant, the neck of a bull, the horns of a stag, the breast of a lion, the belly of a scorpion, the wings of an eagle, the legs of a camel, the feet of an ostrich and the tail of a serpent."
(See " The More Destructive Locusts of America, north of Mexico," by Lawrence Bruner. Issued by Professor Riley, 1893).

THE VINE AND PHYLLOXERA IN CALIFORNIA.

That all the American, Australian and other vines are but prolongations of the original Adamic stalk, and therefore subject to kindred ailments from similar environments and general conditions, is well shown by Professor George Husmann of Chiles Valley, Napa County, California, in his earliest work on " Grape Culture and Wine Making in California," published in November of 1883, from which I now quote :—

" It is well known, he wrote, that the earliest beginnings (in California) were made by the Jesuit Fathers of San Gabriel, with what has since become known as the Mission or as it is erroneously called by many, the California grape. It is no doubt a true *vinefera*, whether, as some believe, it was grown from the seed or from cuttings imported from Spain, it certainly bears no resemblance to our native wild vine *vitis Californica*. A few enterprising men saw in its success there the probabilities of a valuable industry. Their experiments were rewarded with abundant crops which even surpassed their expectations, as our " (then)" dry and equable summers favoured the development of the grape, and although it was thought in those days imperatively necessary to irrigate the vines, they found that the Mission always ripened its fruit and would produce large crops, under a very simple and convenient system of prunning, and make a fair drinkable wine in most seasons . . . Many progressive men, encouraged by the evident success with the Mission grape, imported cuttings of choice varieties for trial from France, the Rhine and Spain, often at heavy expense and risk; they were planted in different sections, and mostly found to succeed well. . . . Farmers *found that the lands they had cropped for cereals, until they were exhausted and would not produce grain,* would still yield large crops of grapes for which they had a ready market at home. It is certainly not surprising if they became over sanguine until everybody and their neighbor planted grapes. As the Mission was known to be productive and they could sell all they could grow (the communistic organization at the Mission; their planting of carefully selected cuttings and rooted vines in rich virgin soil, and the main object for which they originally laboured being to produce the choicest wines for personal use, with their special facilities for enriching their grounds and for systematically prunning the vines after the most approved continental fashion, as also the snug sheltering tree fringes surrounding their establishment, were evidently completely overlooked by neighboring vine-planting farmers, and hence doubtless—the origin of existing phylloxera trouble). A good many vineyards of this (mission) variety were again planted together with a large acreage of Zinfandel and Malvasia. The vineyards were to a large extent planted by men who had little appreciation of fine quality, but planted grapes *simply for the money they could make out of them"* . . . (quite regardless of after consequences).

" Another mistake which many of our planters have made, is the persistence with which they have planted, and are planting even now, the vinifera cuttings and vines, *in districts affected and nearly destroyed by the Phylloxera.* They ought to profit by the lessons taught in France, and all over Europe, by the devastated vineyards which have reduced the crop of France to about one-third of what it was formerly, until the greatest grape growing nation on the face of the globe cannot raise sufficient for her own consumption and has to buy from all her neighbors to meet the demand of her customers. The devastations made in our own vineyards should have convinced the most skeptical." (The italics are mine).

The alarming official report which was published at the commencement of this year by the California State Board of Viticulture, concerning the dreadful condition of vineyards through the Napa county—where Professor Husmann's vineyard is—should favourably commend his, the professor's good judgment, though heretofore despised by local vineyardists, as the following extract from said report will show: " Every vineyard portion of Napa county has been visited and inspected. . . Since my last report, two years ago, vineyards in this county have been greatly lessened in number and in area, in many portions of the county. Commencing ten years ago in the lower end of Napa Valley, and supposed to have been brought from Sonoma Valley, the phylloxera has spread almost the entire length of the valley in the direction of the prevailing wind. Two years ago a few vineyards in the Napa District and some in the Yountville District were infested. Since that time it has spread with great rapidity. In many cases vineyards of considerable extent have, in the meantime, almost or wholly disappeared. This will account for the smaller number of vineyards reported this year.

" No remedy to prevent the spread of the disease has been discovered . . . In almost every vineyard visited, where the phylloxera has made any headway, the vines were allowed to stand without treatment, the disease taking its course. When the vines were dead or nearly so, they were pulled out." During my visit through the Napa Valley in April of 1892—accompanied by Professor Husmann, I was surprised to find nearly all the vineyards most unreasonably crowded with vines, in many, they ranged from three to six feet apart. Amongst the few exceptions was one owned by Judge John A. Stanly, area 125 acres, and as a result he receives most excellent returns. In addition to the commendable manner he has adopted in the planting arrangements, the vineyard is snugly surrounded by fringes of eucalyptus globulus, and in his report to the viticultural commission, from which I now quote, he wrote:—"Since I planted my first resistants, within three miles of my vineyard, 500 acres have been planted to vines and eaten up by phylloxera. My vineyard is flourishing." The Judge assured me that the eucalypti fringes thoroughly protect his vineyard from hoar frosts and severe wind storms. Since the said official report was issued, the phylloxera plague has been discovered in Southern California, and of which the San Francisco *Examiner* of April 1st, 1893, writes as follows:—"The phylloxera, the insect pest which has caused such a great loss in the vineyards of Napa and Sonoma counties and which has destroyed hundreds of thousands of acres of vines in France, has appeared in Southern California in a small vineyard near Santa Ana. Coming after the immense damage that has been done by the Anaheim disease, (a new scourage) this is most discouraging to the vineyardists of this section of the State."

Professor Husmann's statements regarding the origin and early history of Californian grape culture—apart from any other cause in eastern States, furnishes colorable reasons to justify the following opinion published in a German work on the Phylloxera by Dr. Geo. David:

"The nature of the (phylloxera) scourge having been determined, speculation became rife as to the cause of its appearance. It was at first thought that the phylloxera was indigenous to Europe, and that certain external influences had brought about its sudden and extraordinary development. This view, however, soon met with contradiction, *and it is now proved beyond the possibility of a doubt that the disease was directly imported from America.* At Fonelbe, near Tarascon; at Florida, near Bordeaux; at Klosternenberg and Oporto, in fact, in whatever point the disease was first discovered, it is distinctly traceable to the introduction of American plants The phylloxera of Europe and of America is identical; this is proved by the researches of Mr. Riley, who has published a comprehensive work upon the subject. In America, to the east of the Rocky Mountains, the phylloxera abounds, and no European vines have there been successfully cultivated. In those regions, however, a long struggle with the insect has resulted in the development of varieties which are capable of resisting, *to a certain extent,* its ravages, but which are in all other respects inferior to European vines. In Europe the case is different; and the vines attacked by an enemy that they had no hereditary tendency to resist, have inevitably succumbed."

FRENCH REPORT ON THE BI-SULPHATE OF CARBON TREATMENT FOR THE *Phylloxera vastatrix* SCOURGE; translated and re-published by Mr. A. K. Finlay of Glenoriston, Australia, September, 1880, folios 8-9—Said report contains the following concluding remarks:—

"The first applications of bi-sulphide of carbon were made between the 15th of March and 4th of April, 1877. The lower part received a repeated treatment at the rate of fifteen drachms per square yard, while in the upper portion twelve drachms per square yard was applied in a single operation. In 1878 a repeated treatment was carried out between the 15th and 30th of May, over the whole vineyard. On the 15th of July some insects still showed themselves on the lower rows, and an injection of seven drachms per square yard was applied in a single operation. In 1878 a repeated treatment was carried out between the 15th and 30th of May over the whole vineyard. On the 15th of July some insects still showed themselves on the lower rows, and an injection of seven drachms per square yard was given to the three first lines."

Turning to folio 9 of said translation I find the following interesting paragraph:

"In order to understand the wonderful rapidity with which the phylloxera advances, it is necessary to know something of its character. The insect is akin to the aphis tribe, whose fecundity is due to its strange system of generation, which has been thus succinctly described: "The congregations of the aphides consist, in spring and summer, of apterous individuals, and of nymphæ with undeveloped wings. All these are females, which give birth to living young, *sans accomplement preailable*. The males are produced towards the end of summer or during the autumnal season. They fecundate the last broods produced by the females first mentioned, which broods differ from their progenitors in requiring impregnation prior to the continuance of their kind. They lay eggs after the sexual intercourse, and these eggs produce, in spring, the broods above alluded to, which are capable of producing living young without assistance from each other." Article on Entomology, in "Encyclopaedia Britannica"—"Towards the end of summer a winged generation appears, the migration of which form one of the most rapid means of spreading. They are produced upon *the decomposition* of rootlets which have been during the summer subjected to attack. Opinions differ as to the sex of these winged insects. M. Balbiani, one of the commissioners appointed by the French Academy of Sciences to investigate the subject, hold them to be females, while Mr. Riley asserts that they are both male and female. This seems, however, a matter of small importance, since there can be no question that they propagate the plague" (?) (This is a popular delusion as they are simply an affinitised form of nature's scavengers evolved from their torpid condition by the agency of our soil, empoverishing greed for personal gain). The males having served their purpose of fecundating, die; and the females, having laid the impregnated, or as they are called, "winter eggs," enter upon a state of hybernation, to recommence their work of generation in the spring. The winter egg, likewise, in the spring fulfils its functions and produces the nymph which brings forth living young, as already described. It is considered that eight generations of these nymphs are produced during the season, before the appearance of the veritable female; and it has been calculated that the progeny of a single insect may reach, in a year, the number of 5,904,000,000. These figures are held to be too small by some statisticians, but they may be very liberally reduced, and still show the overwhelming forces which the enemy can bring into the field. Before such numbers, the most careful organized system of defence must receive many a rude shock.

"The full grown phylloxera is a small almond-shaped insect, about 150th of an inch long and 130th of an inch broad; it is armed with a powerful proboscis, or sucking tube, with which it pierces the roots of the vine. Although it may be said generally, that the phylloxera is a root-attaching insect, it is also occasionally found in galls upon leaves, more especially in the case of American varieties; and it has been proved by experiments that, while differing slightly in appearance, the root and gall insects are virtually identical, and that each is capable of taking the place of the other. The winged insect already alluded to is the shape in which the phylloxera is most to be dreaded. The wings are comparatively large, and the insect, being light, (resembling fine thistle-down) it is carried great distances by the wind, passing easily over rivers, *forests*, and long intervening spaces which are not planted with vines. That this forms the principal means of migration is proved by the spread of the pest being generally in the direction of prevailing winds. The phylloxera has, however, other means of locomotion, and, *in favorable localities*, marches with great rapidity. There is no sort of soil, with the exception of sand, which has yet been found to hinder its progress; but level, open country, more especially when it is of a nature *which cracks with drought* is the most favorable to its advance. *Atmospheric conditions have their influence and very wet seasons have always shown a diminution of* the scourage. Still no permanent remedy can be looked for even from the most continuous rains."

THE PHYLLOXERA IN AUSTRALIA.

The first report of the presence of phylloxera in Australia came from the Geelong district, in 1878, "(where as in California "*farmers found that the lands they had cropped for cereals until they were exhausted, and would not produce grain* would still yield large crops of grapes.")" "According to the Government inspectors, Messrs. Wallis and Hopton, thirteen vineyards were found to be infected with the disease, and were uprooted and destroyed. The infection, in some cases, had evidently been of long standing, and the vineyards throughout the district generally presented a most delapidated appearance. Unfortunately, the inspectors were appointed too late for any beneficial measures to be taken, for the flight of the winged insects was already nearly ended. The inspectors therefore confined their report to certain observations and recommendations. They expressed their opinion, aggreeing with experience elsewhere, *that the disease was not due to poverty or neglect* (!!) but that it had been imported into the Geelong district, and spread either through the agency of the winged female, or the distribution of rooted vines or vine cuttings. They ascertained that the flight of the winged insects commences about the end of December, and continued till the first week in February—an observation which is of value in considering the work of eradication. The inspectors concluded their report by recommending the adoption of prohibiting measures in regard to the importation of rooted vines and vine-cuttings and the uprooting of all *uncultivated* vine-yards, at the expense of the proprietors.

"During December, 1879, by direction of the chief Secretary, some experiments with bi-sulphide of Carbon were made. The more immediate object of the work was to test, as nearly as circumstances would admit, the efficacy of bi-sulphide to stay the spread of phylloxera, when used in the manner patented by Mr. Rohart, of Paris. The Rohart method, in brief, is to bury, at convenient distances and depths, around the roots of the diseased vines, cubes of porous wood previously impregnated with bi-sulphide of Carbon, and coated with a varnish the composition and manner of application of which is a secret reserved by the patentee. The experiments were devised and conducted by Mr. Manly Hopwood, Chemist and Analyst to the Department of Agriculture. At the time they were carried out the country was very dry and the season much advanced, circumstances which militated greatly against the success of the trials. In lieu of the patent cubes of M. Rohart, Mr. Hopwood used small chip boxes, filled with dry sand, and cubes of "Infusorial" earth, baked to expel moisture. Both these substitutes were charged with bi-sulphide of carbon at the moment of use, and buried around the vines in holes made with an iron bar to the depth of from ten to twelve inches. By experiment in the laboratory it had been found that the evolution of bi-sulphide of carbon vapor from baked infusorial earth is comparatively slow and even.

"Mr. Hopwood's experiments yielded results agreeing in the main with those obtained in Europe, and led him to the conclusion that the employment of the Rohart remedy early in the season would probably be efficacious in destroying the underground phylloxera. But since the system is only applicable before the insects have reached the surface in the winged state, he points out the great importance of determining the proper time for the application, and rightly insists upon the necessity of early treatment. It is to be regretted that the experiments made by Government do not appear to have stimulated private enterprise in the same direction; for, so far as can be ascertained, no further trials of insecticides have been made in the infected districts." (The real cause remained untouched.)

Notwithstanding the Victorian Inspectors having agreed to believe that the phylloxera *"disease"* in and around Geelong *was not due to poverty and neglect"* the actual facts of the case told a very different tale, as it was well known to many local residents that a similar process of vineyard forming obtained in Geelong to what happened—as stated by Prof. George Hussmann, in California. To my personal knowledge the Geelong vineyard plots and much of the vineyard soil in other parts of Victoria, Australia would only yield rank weeds such as thistles and sorrel from a similar soil impoverishing cause when planted with vines.

PROFESSOR F. W. MORSE

In a work entitled "Observations on the Life, History and Habits of the *Phylloxera in California*," made from 1881 to 1886 by Prof. F. W. Morse, assistant in the United States General Agricultural Laboratory, that dreadful insect's mode of attack is described as follows:—

"There is one point worthy of note as throwing some light upon the resisting power of vines; it is the manner of the insects attack. In the common vinefera even, they show preference for particular spots on the roots, selecting those places where the bark is softest, usually near a crack. From this they extend upward and downward along the line where the tissue is continuous from that spot; and scarcely ever do we find them working at right angles from this line. When the sap begins to ooze out and rotting sets in, they precede it closely, always leaving a number of insects to continue the destruction until the spot becomes completely rotted and gives out no more sap. Large numbers of insects will often be found feeding upon such spots, apparently reluctant to leave them·as long as any sustenance can be derived therefrom. So closely is this mode of working followed, that in many old Mission vines they will be found only on a single spot, while the remainder of the root is free from them. A root covered with a fuzzy bark is noticeably objectionable to them, a harder one with cracked or loosened bark is preferred.'' (Precisely the mode of attack in Australia). '' Upon a thorough resistant (?) stalk the insects act quite differently. They are usually scattered about apparently at a loss to know just where to begin operations. Their first piercing are made, and instead of a deep rotting which completely kills the bark to the woody tissue, a slight thin blackening of the bark takes place, which does not extend further, and if made on the firmer rootlets, will often peel of, leaving the root perfectly smooth. (Decomposition of vines must precede an effective attack of insects.}

CONCERNING KNOWN REMEDIES.

"I abstain purposely from description of any chemical remedies, because I believe them too costly and at the same time not effectual enough. They give us no guarantee—even if they could be so thoroughly applied as to exterminate all the insects of permanent security; as they may at any time be again transmitted to the same vineyard, making continued applications necessary, generally with great danger to the vines. Only in cases when it is desirable to save a valuable piece of vineyard of a choice variety, it may be advisable to use Dr. Baner's mercurial remedy, which, so far, is the most promising, least dangerous (?) and cheapest of all that have been tried. Insecticides, of whatever kind and description, are too costly in their application, and have to be renewed too often to ever become practically applicable here or even in Europe. The lowest cost of their application, of which I have seen an estimate, is about thirty dollars per acre, more than the general annual cost of cultivation, and this is only a temporary remedy which must be renewed every year to be of any use at all. Besides great care must be exercised in their application, for an overdose will kill or fatally injure the vines. The pest is liable to appear at any time, and thus it needs constant doctoring '' (as do the human plant)'' with costly remedies to keep the patient even in a state between life and death.''

Professor Husmann in his subsequent work on "Grape Culture and Wine Making in California," published on October 20th, 1887, referring to the Phylloxera and the much relied on "resistant" says (folio 83):

"First of all it is necessary to dispel the illusion entertained by some that resistant vines are such as are not attacked by the Phylloxera. So far as our knowledge extends at this time, the insect will feed on any and all of the members of the true vine tribe (vitis proper) *when occasion offers ;* but it is evident that some are better adapted to the taste *or nature of the phylloxera* than others, and are therefore more numerously infested when planted in the same ground with others ; just as cattle will pasture on the sweet grasses in preference to the sour ones. The European vine (vinefera) appears on the whole to be the one most uniformly adapted to the insects taste in all its varieties, and is always attacked in preference. It evidently offers the best conditions to the life and multiplication of the pest. It is not, then, a proof of non-resistance when a vine is found to be more or less infested ; for, as far as we know, there are no true vines of which the *phylloxera* will not attack the roots when presented to them. . . ' . . . .

But every vine, like other plants, is subject to certain conditions of *soil, climate and atmosphere for its welfare.* Any vine or any other plant may be planted where from unfavorable conditions it will not flourish, and where a slight addition to the adverse influences may cause it to either die or maintain only a feeble existence. . . . . The resistant vines are no exception to this general **rule.** They have been planted and expected to yield satisfactory results, where vines have been fruited for twenty or thirty years without the use of a particle of manure, and where, as a result, the old vines as well as the new "resistant" ones have died from sheer inanition. They have been planted where no vine ever should be. (And hence the existing phylloxera trouble).

THE PHYLLOXERA QUESTION.

"That this is a serious one, likely to effect our industry in all its branches, will hardly be denied by anyone. If we look at the devastated vineyards in Europe, if we consider the ruin it has brought to thousands of formerly happy and contented homes in France, how its ravages have decimated this leading industry, so that now they do not produce wine enough for their own consumption, but buy where they formerly almost supplied the world ; how its ravages are already felt in Algiers, in Austria, and wherever vines are grown—we will hardly question that it is the great disaster threatening everywhere, including this continent. Indeed, we have evidence sufficient of its destructiveness in this State, it will make itself seen and felt, and no mechanical or chemical means have as yet been found that are of real practical value. All the insecticides that have so far been tried have proved too costly and impractical in their application."

COMPARATIVE RESULTS.

Since the issue of **Prof. Husmann's** first work from which I have also quoted, adverse climatic conditions have been steadily increasing throughout America from the cause already stated. In proof of which, extracts from the respective works in parallel positions, with regard to California somewhat testifies :—

"American Grape Growing and Wine Making," November 9th, 1883.	"American Grape Growing and Wine Making," November 20th, 1887.
" A visit to this shore in the summer of 1881 convinced me that this was the true home of the grape, and that California with her sunny and dry summers and her mild winters, was destined to be the wine land of the world; that promised land where everyone could sit under his own vine and fig tree. *Disease* of the vine are here comparatively unknown, the rainless summers, when no showers are expected from May until September, allow nearly all the crop to ripen every year. . . . These favorable climatic conditions simplify the culture and training of the vine, the gathering of the fruit, and the operations in wine making. . . . In this climate it becomes possible that one man can own and superintend hundreds of acres of vineyard."	Apologizing for the delay in the issuing of his new work, Prof. Husmann wrote : " I hoped to complete it before the vintage so that it could be of some use perhaps during its progress. But unavoidable delays have drawn it out to the end of the vintage of this truly abnormal year, abnormal in its late and destructive frosts, its hot winds during summer causing a great deal of coulure, and its unusually hot weather during the vintage, It has been one of the most difficult seasons to handle a vineyard and wine cellar which will ever occur here I trust, and has taught us many and severe lessons."

Now, in 1893, Californian vineyards are rapidly becoming hopelessly diseased.

VINE TROUBLES IN FRANCE.

(From the San Francisco Chronicle of December 7th, 1892.)

PARIS, November 15.—The annual vintage, one of the greatest fetes known in French rural districts, is ended ; the grapes are gathered, and the vignerons, their work finished, have filled their glasses and sung, for the last time until another autumn, the old refrain :

> Bon Francais, quand je vois mou verre
> Plein deuce vin couleur de feu,
> Je songe en remerciant Dieu
> Qu'ils n'en ont pas en Angleterre !

"All the slopes in the south, the center, the east, and the west, all those in the Burgundy and in the Gueyenne, in the Champagne and in the Gascogny, in the Lorraine and in the Languedoc; all those plains of Anjou and Touraine, as well as the mountains of Auvergne and of Dauphiny, have been heard from, and we know now of the quantity and quality of this year's wine harvest.

"For a quarter of a century the phylloxera has been among French vines. The Gard, the Herault and the valley of the Rhone were the regions in which that destructive insect was first found, and then came the turn of the Bordelais and the Charente. The plantations in the center of France and those in Burgundy were not touched but in the afflicted districts milldew, odium, and other evils also fell on the great fields. The average produce of wine fell from thirty-five and forty hectoliters per hectare to twelve and fourteen only, and in four or five years land had lost $200,000,000 of its value.

"France is not the largest wine producer in the world, however. There are 21,215,125 acres of vineyards in Europe, and all the rest of the earth shows less than 1,000,000 acres Italy figures at the top with 8,010,000 acres planted in grapes, France comes next with 5,832,000 acres, and Spain is third with 3,745,000 acres. Last year Italy produced 806,000,000 gallons of wine, Spain 702,000,000 gallons, Austria-Hungary 235,000,000 gallons, Germany 63,800,000 gallons; Switzerland 25,800,000 gallons, and over 70,000,000 gallons were produced in Algeria.   .   .   .   .

" Phylloxera in Champagne ! You can perhaps imagine the excitement there was all over this country, and the excitement was not lessened the past summer when it was discovered that other vineyards had been attacked. Champagne making is the fortune of that part of France, and it is an enormous fortune. In plentiful years the production of white champagne wine is not less than 80,000,000 bottles. But the vineyards of Champagne are more vigorous and healthy than all others, hence they were able to resist the phylloxera. Still tons of sulphide of carbon have been used up there, and the vines that were infected with the disease were cut away and burned and the soil was poisoned.   .   .   .   .

"There is not much possibility that the prize of $60,000 offered by the State to the discoverer of an efficacious means of destroying phylloxera will ever be awarded."

Just as certain as the suicidal practice of depleting the human body by indiscriminate blood-letting to cure disease had to give way to a more rational mode of treatment, so also must the prevailing theories of vine cure succumb to the light of wisdom and common sense.

The phylloxera made its appearance in the south of France about the year 1863 and destroyed a great many vineyards before its presence was discovered. One of the first victim's whose vineyard was destroyed and rooted up in 1867, replanted another part of his estate the following winter. He divided the ground to be planted in three different parts of equal size, and as much as possible, containing the same nature of soil. Each plot was planted with the same kind of vine, and received the same treatment even after, with the only difference that each lot of vines was planted on a different scale. One was planted 4 x 4, or four feet every way, the other 6 x 6, or six feet every way, the last one planted 10 x 10, or ten feet every way.   The whole vineyard was treated alike for thirteen years after-wards, well worked and heavily manured periodically.   In 1880, a com-mission from the French Vinegrowers Association was sent to that district to inspect the vineyards in connection with the phylloxera.   The trans-lation of the report relative to the above mentioned vineyard runs as fol-lows:—

" When we arrived at the estate of Mr. X, in the Department du Gard, and had been introduced by Mr. B, Mr. X received us in a most kindly manner, showed us all round the vineyard, kindly supplying us with any information we required. This vineyard, which is divided into three equal parts, has been planted about thirteen years, after having been destroyed by phylloxera in 1867.   Each part was planted on a different scale.   One was planted 4 x 4, another 6 x 6, and the third 10 x 10.   Each of them has been well worked, well and regularly manured every three years ever since they have been planted.   The vines in the first part planted 4 x 4, are nearly all dead, except a few around the rows *along the roads*; those in the second, planted 6 x 6 are still all alive, but look very sickly, and Mr. X told us that the grapes he had gathered from them for the last few years did not pay his ex-penses, and if it had not been for the hope he still had of curing them he should have rooted them up long ago.   The vines in the space, planted 10 x 10 were grow-ing luxuriantly, and were bending under the weight of their fruit.   With the help of Mr. X, we dug up, ourselves, vines roots in several places in the three different parts of the vineyard and we found that they were pretty well all alike, covered with insects.   But roots from the wide plantation did not seem to be affected by their presence, while the roots from the two other parts were all more or less advanced in decomposition."—(Where, when and how to plant the vine, by I. Couslandt, Australia, 1883, folios 13 and 14.)

It should here be noticed that though the above vineyard soil was over-run with phylloxera—evolved from the closely planted vines, they were utterly powerless to penetrate those planted 10 x 10.

ATMOSPHERIC GERMS.

The following extract from a seriously interesting lecture delivered at the Cooper Medical College, San Francisco, on the evening of Friday, February 24th, 1893, by Dr. W. F. Cheney, faintly illustrates some of the " blessings" we enjoy from our chronic contempt for the earth's forest-lungs in and around centers of population :—

" The lecture proved quite instructive and was listened to attentively by a large and appreciative audience."   The lecturer said :   " To prove the presence in the air of living forms this simple test will suffice :   Take a little glassful of buillon broth, filtered to remove all bits of beef fibre.   Such a liquid is clear as crystal and shows under the microscope not a sign of life, no matter how closely exam-ined.   Let this stand for a few days in a warm place and then examine it again. The fluid is no longer clear, but turbid, and a drop placed beneath the microscope will be found crowded with living things that jostle one another and hasten to and fro across the scene like busy human beings in a city street.   We scoff, perchance, at miracles.   Yet here is surely one.   What magic art has wrought this change? Whence have come these innumerable specks of life?   This was an unsolved problem long after the microscope had revealed their existence.   Men could not doubt their eyes—the living things were present in the fluid beyond question, but what had been their origin ?

(Wholesale forest destruction and heartless selfishness in many such like forms as the following extract from the *S. F. Examiner* of April 26th, 1893, denotes: " The Grand Jury will perform an important public ser-vice if it gets to the bottom of the frauds in sewer and street paving con-tracts of the last few years.   The city and the property owners have been badly swindled by conscienceless contractors.   Contracts have been violated, specifications ignored, and money collected for work that has not been done.   The sewer commission has found numerous instances in which every rule of good work has been violated.   Rotten brick has been used, mortar has been made of mud, and the trenches have been dug so that the sewer in the center of a block has been higher than at either end.   This is something worse than swindling.   It is a crime that includes the deliberate poisoning of the people who live along the line of the rotten work.   This form of rascality is deserving of severer punishment than falls to the lot of the common thief or pickpocket.   If the man who breaks into a house to steal a loaf of bread gets a dozen years in prison, the rascal who cheats the citizens and the city by laying a rotten sewer deserves a lifetime behind the walls of Folsom.")

"It was discovered that the generation of these swarms can be prevented in a very simple way.   That if the bullion is kept for some time at boiling temperature and then sealed at once to prevent the access of air, it will remain clear for days, for weeks, for indefinite periods, and show no trace of living forms on microscopical examination.   The sole difference in the treatment of the fluid in the two cases has been the exclusion of air.   And so men came to look in the air for the source of the changes so long misunderstood.   Then slowly, step by step, came the name *cocci*, from the Greek word meaning a berry or pill.   Others look like short rods, and hence get the name *bacteria*, from the Greek word meaning a rod.   Still others of a similar shape, but a little longer, are called *bacilli*, from the Latin word meaning a staff.   And, lastly, there are forms called *spirilli*, because they look like long, twisted spirals.   Though science gives these distinctive names to different forms, yet popular usage designates all germs as bacteria, regardless of their shape.   In size, though all are visible to the naked eye, there is nevertheless great variation among them.   Some of the *cocci* are but 1–20,000 of an inch in diameter, so that 20,000 of them together in a heap would occupy only one cubic inch of space.   Some of the *spirilli*, on the other hand, attain a length of 1–300 of an inch, but with germs as with men, the stature of the individual bears no relation to its influence or to its possibilities.

"In structure all these living germs are but simple vegetable cells.   Imagine a minute speck of matter of jelly-like consistency, the surface a little more firm and dense than the center.   Such is the living cell, the last step in the subdivision of all living structures.   Tho anatomist, with his knife, can dissect the animal body and reveal the different tissues of what it is composed—the muscles, the bones, the arteries and the nerves and the connecting strands that binds the parts together.   But with his microscope he goes on far beyond that point and learns that each muscle, bone and nerve is itself made up of multitudes of individual cells, differing vastly in shape and s ze according to the tissue in which they are found, but all essentially the same in structure.   So the botanist can take apart the plant and show its various tissues, but each of those tissues, beneath the microscope, becomes at last a mass of tiny cells.   Beyond this point, either in animal or vegetable tissue, no eye has ever seen, and these cells become knowledge that in the atmosphere everywhere float seeds or germs of life, constantly on the alert for a congenial spot where they may alight and make their home.   With this discovery the old doctrine of spontaneous generation had at last to be discarded.   It was not given up without a struggle and many a wordy battle, but to-day there are few indeed who believe in it.

"Not content with finding that the earth contains such germs, science has devised ways for estimating their number ; has made out their structure; has learned how they reproduce their kind, and has become familiar with the habits of individual varieties.   An ingenious French scientist, after painstaking experiments, has calculated that in the ordinary atmosphere of a large city there are 2,000 germs to every cubic yard ; in the air of a room or house in winter, kept closed to exclude the cold, he estimates that there are 45,000 germs to every cubic yard, and in the wards of a long-used hospital he found 90,000 germs in the same air space.

"The same French scientist found that in summer, whether in city or country, the air contains three or four times more germs than in winter ; that the air at elevations is always less densely populated by them than that near the surface of the earth, and that the air on sea or on mountain heights is almost entirely free from them. It is generally to be regretted that this scientist did not extend his investigations to the atmosphere of one of our Mission-street horse-cars, habitually crowded with passengers and with every door and window tightly closed. The man who does examine one of these for germs will, no doubt, find "there's millions in it !"

"The dwellers of the air vary greatly in size. Some appear under the microscope as minute, rounded dots. These have been, therefore, the points of beginning for all living things. From such an assumed origin comes every plant, every animal and every human creature, by multiplication and differentiation, by growth and development. The difference between a bacterium and a man is that the bacterium never gets beyond the starting-point. It begins and ends its existence as a single cell. Such as it is in structure all living creatures once were, and the mystery of its origin and creation is as great as that of man. The humble coccus, standing on the very threshold of that domain where living things abound, presents a problem as profound as that of the genesis of a human soul. It possesses *life*, and that possession in spite of its simple form makes the coccus an object of absorbing interest and of deepest awe.

"One of the most familiar manifestations of the presence in air of unseen germs is the formation of mold, such as is frequently seen on bread, fruit or cheese. To the naked eye this is only an ugly scum or fuzz that can be easily scraped away from the surface and causes no perceptible changes in the parts beneath. Under the microscope this would become a luxuriant field of green as beautiful as the expanse of waving grain that clothes the plains in spring-time. Air always contains these mold germs, but they do not take root and grow on the surface where they chance to light, else mold would be the rule instead of an occasional occurrence. These infinitisimal seeds must be well watered or they die. This we express when we say that mold forms only in places that are moist or damp. The mold germ that thrives on bread is not the one that finds nourishment in the juice of fruits, nor the one that lives by preference in the barnyard.

"Some years ago, during excavations in the buried city of Pompeii there were found several jars of preserved figs, hermetically sealed, which had been prepared by some good housewife of that ill-fated city, eighteen hundred years before When these jars were opened the figs proved to be as fresh and delicious as if put away only the summer previous. It is said that that discovery taught the present century the art of preserving fruit. This art, so familiar to every household, depends merely on the exclusion of the germs of fermentation. There are other kinds of germs whose importance is very great, for they find their residence, not in the dead, but in living tissues, and thus become a menace to the health and even the life of man. Does all air contain disease germs, and is there any way to prevent their presence there ? These are questions of vital importance to the human race. Fortunately for man, such germs are *not* always hovering about him. Air is seldom entirely free from microscopic forms of life, but the majority of such is perfectly harmless to man. (?) It is only exceptionally that air contains those varieties that cause disease. There are two sources for their occurrence in the air of any particular locality :

"First: They come from decomposing material; for while some disease germs will not, so far as known, multiply outside of the body of infected animals or human lungs, there are others that do multiply abundantly under favorable conditions as to food and temperature. This is the case with germs of typhoid fever, diphtheria, glanders, cholera and erysipelas. All of these thrive in filth of any sort. The second source of disease germs in the air of a locality is the presence of cases of disease. From a human body in which these parasites are living some are constantly escaping. Deprive disease germs of the inefficient sewers, which are to them as the promised land, flowing with milk and honey and build more harbors of refuge, equipped with the means to put the enemy to route—then and not till then will the city become a safe place for residence when epidemics threaten." (*S. F. Bulletin, February* 25, 1893.)

THE MEDITERRANEAN FLOUR-MOTH.
(*From the San Francisco Morning Call, Dec. 8, 1892.*)

"The terrible ephestia kuhniella (Zeller) is spreading millions of darkening wings, and coming as the locusts came by the Nile. They started from Eastern Europe several years ago, and, advancing like the cholera, and by similar means, have now reached the Pacific Coast. They are stopping whirring flouring-mills, and are in our daily bread, our morning gems, pancakes and mush and our evening pies and sweet cakes. There is no joke about it. The Mediterranean flour moth is really doing those things. It has just gained a fair foothold here, and, as it rapidly spreads, it is proving a very expensive, annoying and disagreeable insect pest. The annoyance promises to increase rapidly, and a determined warfare will soon be waged against it generally throughout the State. The flour-moth had not been heard of here until two or three years r go when it appeared in a few mills in this city, and it has been but a few months since it began to attract general attention among the millers, as it gradually spread from mill to mill, but it has already cost them thousands of dollars, and they have just become thoroughly wakened to an alarmed inquiry about the pest and how to fight it. It seems strange that a noiseless little moth less than an inch long and so light it cannot withstand a breath should stop the steam driven shafts of a great flouring-mill from turning, but it does. It does more than that, too. It ruins fresh flour and meal by the barrel, and it makes the housewife turn up her nose in disgust while she is getting breakfast and then go back to the groceryman with a complaint. It makes the groceryman complain to the miller, and along the line from the miller to the breakfast table there is turning up of noses, a loss of good flour and oatmeal and buckwheat and a general wondering what is the matter. People may now understand it all and lay the blame on Providence or foreign immigration. (Yes, "lay the blame on Providence," as we have always done, to, if possible, shirk our own responsibilities.)   .   .   .
It seemed odd that an insect at this late age of the world should discover for the first time that flour was good eating and that flourmills made good homes after letting flourmills alone for ages, but Professor Johnson could not explain this late progressive move of the flour-moth."

A WAIL FROM MALTA.

A highly interesting letter from the pen of **Mr. John H. Cook, of St. Julian, Malta**, dated September 22nd, 1892, **referring to " The remarkable falling off in the quantity and quality of the fruit"** in that part of the world, appeared in the printed Reports of the **Consuls** of the United States, No. 149, February, 1893, (folios 261–263). **Mr.** Cook declares, and evidently U. S. Consul Worthington affirms:—

" **That** unless some energetic measures be adopted to counteract the influences of the causes that are at work, within a comparatively brief period of time many fruits, such as the orange and mandarine, for the growth of which the Maltese Islands have so long been famous, will be either exterminated or will be cultivated only as curiosities. It is, I believe, a well known fact that the orange, mandarine, lemon, fig, nectarine, olive, apricot, apple, pear, grape—in fact all fruit and most kinds of vegetables grown in these islands are at present affected with a variety of diseases that not only attack the fruit, but also devitalize the tree to such an extent as to jeopardise its very existence. The Malta orange is at the present time affected with four different kinds of diseases.   .   .   .   Another foe, which is of an even more deadly character, is the beetle *cerambyx*, which, while in the caterpillar stage, bores its way into the heart of the trunk of the nespola and the apple trees, and causes them to rapidly decay. But it is not the fruit gardens alone that have been invaded by insect pests. The pea crop of last year was seriously diminished by the attacks of a worm called the " cadell," and of an aphis called the " blanqueta."   .   .   .   The wheat, the sulla, the cotton—the staple productions of the Malta soil—all have insect foes in numbers that are out of all proportion to the area of the districts in which the crops are grown. To what are we to attribute this invasion of insect pests?"

(Comparatively tiny deforested and **therefore** "devitalized" Malta standing about midway between the mighty African Sahara desert, and deforested Europe and Turkey, may be likened unto **a** much enfeebled man who, from force of circumstances is compelled to reside continuously within an imperceptibly diseased atmosphere generated from pestiferous environments, and whose flickering vitality has been sustained to its utmost limit by means of artificial stimulants and legislative enactments. It should now be evident from a careful perusal of this paper that the sensible suggestion herein quoted from the New York *Herald* concerning the reclaimation of Gallilee by the restoration of her forest-lungs—and, **let** me add, her refreshing inland lakes, coupled with a genuine desire **for** our neighbor's welfare, equally applies to the earth as a whole, and especially **to** Malta. **Then,** and not till then, will **disease insect** laden whirlwinds cease to scatter their retributive **hosts over the earth's sur- face,** as also will **the reign of death-dealing** cyclones, **torrential floods,** protracted droughts, **blizzards and** general pestilence **be dethroned.** Just think of the Mediterranean flour-moth pest now in California !)

Though having selected such reports of destructive cyclones as apeared to furnish ample evidence in support of my contention re the costly fruits from wholesale forest destruction, I must quote **a** little further from **an** article which appeared in the S. F. *Chronicle* of April **21st,** 1893, under the heading:—"Fatal work of a Southern Cyclone," which as said article states —"scattered desolation through two counties of southern Mississippi:—"

"Meridian (Miss.), April 20—Clarke and Jasper counties of this State suffered from a cyclone last night at 7 o'clock, more destructive to human life and more serious in its damage to property than the one three weeks ago. It followed almost the track of its predecessor. Over forty people were killed and nearly two hundred were more or less injured. These figures may be increased when all the stricken districts are heard from. The destruction of property will amount to hundreds of thousands. Entire neighborhoods have been literally swept off the face of the earth.

"The cyclones path *was through a vast pine forest, broken here and there by farms.* Huge trees were uprooted **and** carried for long distances. Near the town of Pachuta lived the family of William Parton, consisting of himself, his wife and three children. Their bodies, except that of the youngest child, were picked up over a mile away, mangled and entirely nude. Their brains had been dashed out. The youngest child had a marvelous escape. It was found half a mile from its home early this morning, uninjured and crying piteously **for** its mother.

"William Fisher, his mother, his wife and their **five** children were **blown away,** and search parties have been out all day, but none **of the** bodies has been **recovered.** A child of Sim McGowan was found dead on a tree top a mile from its home. . . Every messenger from the remote districts brings a tale of suffering and death. The wind is still high to-day, and as each cloud has appeared the people have huddled together in **terror.** . . .

"The cyclone came from the southwest and traveled **in** a straight line until it reached Quitman. There its course changed abruptly. Going in a southerly direction for three miles it took another turn eastward." It doubtless followed the deforested pathway.

The *Chronicle* (same issue) had also **the** following dispatches:—

"MILWAUKEE, April 20,—Twenty-one **men** drowned, one vessel ashore and the life-saving crew covered with glory are among the results of last night's terrible storm at this port. The men who were lost were employed on the new inlet tunnel, and lived on a crib, which is located 5000 feet out in the lake (Michigan) off the pumping station at North Point. . . "The gale developed rapidly into one of the severest easterly storms experienced for several seasons. . . The waves were over fifteen feet high and they dashed over the crib with terrific force."

"ST. PAUL, April 20.—Three feet of snow at this time is most unusual in this State, but that amount of snow fell last night and to day. In some parts of Min-

nesota, the average being over a foot, heavy rain preceded the snow . . falling
steadily this afternoon, up to which time fifteen inches had fallen."

SOME CAPITALISTIC AND HYGIENIC CONSEQUENCES.

Apart from the terribly fierce results to local human life and property
from deforestation throughout the once extensive hog and potato raising,
Mississippi Valley, the whole populace of America are now being made to
roughly share in the trouble from lack of those necessary food supplies.
I am assured that until a few years ago Mississippi Valley hog farmers
derived handsome returns by disposing of their live hogs at from four
cents to five cents per pound in large flocks, and that now, because of the
intense cold, cyclones, torrental floods and otherwise general unreliability
of the seasons, the rearing of hogs and root crops have become almost impossible, and in consequence pork curers have to pay from seven cents to
eight cents per pound for the shortage supplies they can procure. I am
also assured that from said cause and a recently developed swine disease
known as "the hog cholera pest" the output from Chicago, Milwaukee
and elsewhere is now annually reduced by millions of carcases. Some
philosophers may suppose that this is but a trifling matter scarcely worth
noticing, but many others will doubtless be disposed to think otherwise,
and especially on reflecting over the fact that sixty-three millions of
American people who have been educated to the almost daily consumption
of pork in some form, besides many millions elsewhere who look to the
United States for their pork supplies, have now to pay at least twenty-five
per cent more than heretofore for that commodity and in all probability
for a much less nutricious article containing possibly, more or less disease
germs. On the new disease known as "hog cholera" the New York *Homestead* of April 20th, 1893, writes as follows:—

"There is no reliable remedy. Prevention is the one course to be pursued . . .
In buying fresh hogs avoid all public stock yards, and railroad cars or other public
conveyances. Buy from country herds where you know there has been no sickness"
—(a now very difficult task)—

And concerning the terrible disease so common to hog flesh, namely
"trichinæ" the S. F. *Examiner* of March 22nd, 1893, writes:—

" There are at the German Hospital in this city, three patients whose complaint
is unique in the experience of most of the local medical practitioners. They are
William Hunjus and John and Walter Nagel, St. Helena farmers, who were brought
from their Napa home to the hospital on Fourteenth and Noe streets, about a week
ago. The three Germans are afflicted with trichiniasis, and it commences to look
as though the trouble would cost Walter Nagel his life. . . Hunius and the
Nagles owe their affliction to pork infested with trichinæ. A few days before they
were brought to the hospital they ate some sausages containing raw pork, and a
microscopical examination of sections of the hog from which the pork was taken has
shown conclusively that the animal contained myriads of trichinæ."

Concerning the potato failure the S. F. *Evening Post* of April 28th makes
the following comments:—

"From present appearances there will be a potato famine in California this year,
or, in any event, high prices will rule, even if a full supply is obtained from outside
quarters. Stocks are lighter than they have been before for many years, and the
advance in value has brought in shipments from the East as far back as Wisconsin.
There are few dealers here who can recall another instance of the kind, although
for years past Utah and Nevada have been drawn upon to help out at times.
The shipment referred to came to hand Wednesday from Waupacca, and as high

as $2.10 was paid for choice lots. They were all of the Burbank variety. Another carload is expected to arrive and a higher rate is bid for them. From all accounts this will be a very bad year for California potatoes, and the crop will amount to little over one-half of that which is usually raised. The prolonged rains interfered with the planting and did not benefit the crop put in early in the year."

The same issue of the *Post* contains the following doleful dispatch:—

POOR CROP OUTLOOK.

"St. Paul, April 27.—The farmers of Minnesota and the Dakotas are pretty nearly discouraged over the outlook for crops. Not an acre of grain has been sown in North Dakota, nor at any point in Minnesota north of St. Cloud, and there is no prospect that any will be sown in the next ten days. It began snowing Wednesday and an average of eighteen inches fell in twenty-four hours. Since that date it has rained almost continuously and yesterday morning it again began snowing and the fall was steady all day all the way west from St. Paul to Dickinson, N. D. In northern Minnesota, around Crookston, Fisher and many other Red river points, thousands of acres have been converted into lakes by overflowing streams, and all talk of putting in a crop is out of the question."

CALIFORNIA'S DEFORESTING CONTRIBUTION.

The **S. F.** *Chronicle* of April 23rd, 1893, amongst its Columbian Exposition articles was the following:—

"The lumber industry is one of the oldest and most important branches of trade in California, and antidates the admission of the State into the Union. The manufacture of lumber in California can be traced back as far as 1838. . . The exports of lumber from San Francisco by sea (alone) to all countries dating from 1870 have been as follows:—

YEARS	FEET	VALUE
1870	13,679,652	$245,216
1871	17,590,854	312,570
1872	16,517,171	309,325
1873	17,415,287	350,024
1874	9,036,799	176,956
1875	10,024,189	202,912
1876	10,781,220	199,894
1877	13,874,327	267,333
1878	14,596,422	289,374
1879	16,501,075	316,485
1880	14,370,796	307,006
1881	18,269,157	393,283
1882	22,094,393	515,974
1883	14,876,396	332,236
1884	20,231,584	489,642
1885	19,266,070	413,935
1886	15,352,649	294,403
1887	15,911,000	428,008
1888	22,535,740	597,230
1889	18,877,570	457,214
1890	19,169,980	448,074
1891	19,931,521	470,345
1892	21,332,560	495,572

When we couple to the above the many millions of feet that have been used during the said periods in and around California for mining operations, building and repairing, domestic consumption, wharfs, jetties, railway sleepers and other woodworks, forest fires, farm clearings, fences and bridges, etc., minus any commensurate replenishing of deforested lands, we should not now be surprised with existing results to climate, etc.

DANGEROUS EXPERIMENTS.

(From the Toronto *Weekly Globe* of March 11th, 1891.)

"Imagine an application of poison spray to a dying man to restore him to a state of primal health !"

OUR FRUIT TREE ENEMIES.

[By B. G.]

(Ancona, March 5th, 1891.)

"Perhaps I may be allowed to say for the immediate information of some who have come to look upon spraying as an essentially fundamental principle of fruit-growing that in reality it is not so. If this was admitted then what would become of an industry of such immense growing importance to the great interests of this great country. Why, it would simply become extinct, and that in less time than it would take to grow the trees. If men were convinced that they would be compelled to supply a continued spray of liquid poison to the whole mass of their fruit trees, or to any considerable part of them, before they could be sure that anything could be gathered from them for their own or others' use, what would they do ? Would they plant fruit trees at all ? Not under ordinary circumstances.

"Just think for a moment what this thing in its minutia means. It means in the first place that every fruitgrower, before he can ever indulge in the fond hope to see beautiful bright red apples or rich, luscious golden plums on those trees, must keep on hand a heavy stock of poisons in the form of London purple or Paris green, articles that should never be seen or thought of in connection with an orchard or a fruit garden, much less to be used in them. It means further, that a heavy and expensive paraphernalia, consisting of implements and machinery of various forms and designs be kept always on hand for the manufacture, preservation and application of this essential liquid poison many times during the season to each and several of those trees. It means, still further, the possession and use of the constant, watchful, scientific eye; the eye, indeed, of the practised expert and the close observer of phenomena to discern just the exact time when to apply those poisons, some for insects and some for fungus, so as to be most effectual and to be ever ready just at the proper moment to apply them. It means, finally, an immense amount of valuable time and of persevering attention and industry, a large outlay of expenses and a display of talent and business tact in this particular line of work that are granted to the favored few.

"But it means, again, far more than all this and something that involves in its conception an immense amount of valuable as well as of invaluable life to the insect and animal. We all know full well, and some of us to our sorrow, how very dangerous these deadly poisons are, not only to handle, but to anything with which they may come in contact. In spraying trees either for fungus or insects it must be done in early summer time when the most plentiful of insects of all classes, both friendly and unfriendly, are on the move in beautiful activity. To hastily conclude that the whole brood of them are necessarily enemies and ought to be poisoned (including bees) is rash, false and ruinous in the extreme to the best interests of this country and its own population; and so, if we go at this business on a systematic scale we shall find it out to our deepest chagrin and regret. It will be quite safe to say that many of our most valued friends, as well as our most virulent enemies, will be thus destroyed by coming in contact with poisoned fruit trees at this early period of the year. (This equally applies to any fumigating process).

"But again, a liberal application of these deadly ingredients to our fruit trees during the early summer means leaving much of their force and deadly power over and about the fruits themselves, some of which sticking undissolved and visible about the cavity, and some considerable of it adhering to the calix of the eye. These doctored fruits coming into the hands of young eager, incautious children, and animals, are quickly devoured holos-bolos and without a thought of danger until a pain seizes them in the stomach and bowels, and the foundation of deadly disease is firmly laid in the system, but the secret potent cause is never once suspected. Who would like willingly and knowingly to supply an article like this so dire in its effect to the young and rising generation, the hope **and confidence** of this country? Spraying fruit trees in orchards and gardens means danger of a probable contact with other plants and fruits in the near neighborhood, greatly endangering their use, but the most common form of danger is to the domestic cow or the sheep that is allowed to graze upon the tender grass or herb found growing in those gardens or **orchards.** This grass is as surely and certainly doctored as these trees themselves are, and the danger of using it is as virulent and deathly. In the face of all these considerations, considerations of immense importance and frought with serious con **sequences to us, we must** certainly conclude that to rely upon the constant use of **tnese deadly poisons for our** fruit delicacies is not, to say the least, a safe thing to do.

"Let us now for a moment consider the case as "it stands in respect to fungus" —(results from atmospheric putridity). "What is fungus? Well, it is in the first place, a sure and certain indication *of disease and impaired vitality, or bad health,* if you please. It is further an indication, and infallible sign of improper treatment and bad management on the part of the orchardist, *or of wet worn-out or unsuitable soils and locations.* But it may most likely of all be a sure and cer tain indication of an absolute worn-out or exhausted variety. This being the case, what is the most obvious and reasonable thing to be done? Run to the poison bar rel? Not by any means for relief will not certainly be found in this line. Imagine an application of poison spray **to** a dying man to restore him to a state of primal health! A far more rational procedure would be **to call in the** physician, seriously consider the whole matter and carefully learn how **the** case stands."

CONCLUSION.

The law of attraction **briefly referred to at** the commencement **of** this **treatise, is** unceasing in its operations **through** the whole realm of nature and especially so with regard to the formation of aireal reservoirs *i. e.* **rain** storing clouds, **by** evaporation at the tropics and elsewhere by the sun's **agency, as also in the** equitable distribution **of** their migratory contents **over the earth's surface** as beneficent fertilizing aids **to** mankind, by the **attractive power of living** forests, and, had we lived **for** each other's un **stinted happiness as was** decreed by the Allwise Creator, instead **of** for **self as** *we* **determined on,** the existing meteorologic troubles would be un known—clouds would **then have** moved over the earth's surface in detached rhymic **order** yielding up their refreshing treasures at the attractive invita tions of venerated forests **to** enrich **the** many intervening plains, which plains by **such means would in** return have yielded abundance of every necessary provender **for** man and beast whilst the forests could readily be made to abound with the choicest of fruits and fragrant flowers. Now, the whole atmospheric machinery **is** "out **of joint**" and hence our increasing troubles. The sun **still** faithfully **performs its** moisture attracting functions, but not so the once **numerous forests** as they **are now** nearly all **exterminated;** leaving the naked earth to generate cyclone and disease **forming air,** whilst the over-loaded clouds reel hither and thither in an **erratic** manner—frequently uniting over **a** more attractive region of the earth, there meteorologically impelled **to** shower down their contents in devastating avalanches, leaving other **parts to** suffer from protracted droughts and consequent famines, etc., **as the following** illustrates:

RUINOUS CONTRASTS.

From the San Francisco Chronicle,
April 30, 1893.

THE MISSISSIPPI RISING.—LANDS COVERED ON THE MISSOURI AND ILLINOIS SIDES.

ALTON, (Ill.), April 29.—The danger line for the stage of water has been passed, and those who have interests at stake are watching the water creep up to and over their possessions. Missouri points are flooded. Unless a fall soon sets in great damage will follow.

QUINCY, (Ill.), April 29.—The most serious hailstorm known for years struck Quincy and this vicinity and did much damage to fruit and other trees. Windows and conservatories were smashed all over town. The river is rising rapidly and a repetition of last year's flood is feared.

HAIL STORM IN ILLINOIS.

ALTON, (Ill.), April 29.—During last night this vicinity was visited by a hailstorm the like of which was never equaled around here. All vegetation was literally torn to pieces. The Missouri, Kansas and Eastern tracks were greatly damaged. The loss will foot up in the thousands.

From the San Francisco Examiner,
April 30, 1893.

[Special to the EXAMINER.]

DALLAS, (Tex.), April 29.—It was learned from passengers on the eastbound train this evening of the destruction by a cyclone of Cisco, in this State, during last night. There are not more than twenty-five or thirty houses left standing, and up to the time the train passed, about 2 o'clock this afternoon, twenty-one bodies had been recovered from the ruins and there were ten or twelve more persons missing.

APPEALS FOR AID.

The following telegram was also received by Mayor Levy from Cisco:

"Cisco has been destroyed by the most destructive cyclone that ever visited Texas. More than four-fifths of the people are without shelter. There were many killed and wounded. Help is needed to bury the dead, take care of the wounded and relieve those who lost everything."

After the cyclone passed much of the wreckage was burned, having caught fire from overturned stoves. It is therefore probable that most of the missing, about a score, have been burned to death or their

MELBOURNE, (Australia), *Argus*, Jan. 7, 1893.—"The weather is very dry and present outlook for the coming vintage is not so promising as previous years, black spot and odium have made their appearance in the vineyards." (a six months drought).

S. F. Chronicle, April 30, 1893.

PANAMA, April 29.—The Government continues with unabated effort to combat the terrible famine that for some time has been raging throughout the Cauca valley and the horrors of which were recently augmented by the eruption of the Sotora volcano and the consequent damming up of the principal rivers of the district.

ODESSA, April 29.—The abnormal weather continues. The winter wheat crop in the southern provinces has been almost destroyed by the cold. Food prices are rising and famine threatens. The Government will probably be compelled to revive the embargo on grain.

BERLIN, April 29.—Farmers are wailing over the lack of rain. The country is baked and unless a change occurs soon crops will be damaged and we shall have a vegetable famine. The seeds now sown are burned up. Other industries are seriously affected The proprietor of a large dye works says the air is so dry he cannot get colors to take. For the same reason workmen in velvet factories around Chefield find the greatest difficulty in cutting silks, which become brittle, owing to the absence of moisture.

S. F. Examiner, April 30, 1893.

[Special to the EXAMINER.]

LONDON, April 29.—The extraordinary weather continues to be the one vital subject of conversation. For fifty-seven days now there has been no appreciable rainfall in and about London. Farmers are complaining that a few more days of drought means ruin for them. All over the continent, from Italy northward, the same cry is going up from vineyard and orchard and farm. The only people not complaining are some of the large dry goods houses. They say the pleasant weather has almost doubled their sales and that the season has been the most profitable known in a score of years, but the price of vegetables all over Europe dead bodies cremated. Many streets are impassable on foot. Physicians estimate that no less than 200 are injured, of whom forty will die. The property loss will exceed $2,000,000.

keeps going up. This heat is just what the cholera microbe wants, and gradually the line of plague is advancing on the great cities. The cholera is practically epidemic in Normandy.

REMEDY.

An early whole-hearted international combine to restore and righteously protect the planet's forests, and thereby reinspire the long estranged harmony of nature to unite in perpetually singing through her manifold works: "Glory to God on high and peace on earth to men of good will."

PERSONAL.

FROM THE HON. MINISTER OF AGRICULTURE, NEW SOUTH WALES, AUSTRALIA.

SYDNEY, Castlereigh Street, January 22nd, 1892.

MR. JAS. McLEAN :

Dear Sir—I am glad to hear that you intend proceeding to the United States on an important mission. The vast experience which you possess regarding our various resources of wealth, combined with your knowledge of the population, will materially assist you on your tour, and if you are encouraged by a reciprocity of feeling, your trip may be profitably utilized and prove a valuable factor in the progress of the colonies. I trust that you shall have an enjoyable and successful voyage, and that you may ere long be again amongst us, manifesting your general interest in the cause of settlement. Yours faithfully,

JAMES N. BRUNKER.

(*From the San Francisco Examiner, April 10, 1892*)

"Under the heading of "Australian Parasites" we noticed on the 3d ultimo, the arrival in San Francisco of Inspector James McLean from Australia, by the steamship *Mariposa*, on his way to Washington, D. C., in order to place a valuable insect pest eradicating discovery before our agricultural authorities. This discovery Mr. McLean by great research made whilst employed as a forest and settlement inspecting officer, which position he held for many years under an Australian Government. It is now our pleasant duty to further notice that Mr. McLean did not confine his researches to the domain of fruit-destroying insect pests entirely, but to the tracing of cause and effect with regard to the laws of health and disease in the "human plant" as well, and especially relating to the terrible death-dealing disease known as phthisis or consumption. Mr. McLean, who studied medicine in Scotland during his early years, was busily experimenting with a preparation from his insect-destroying specific on sundry consumptives, with marvelous results, in Australia, contemporaneously with Dr. Koch of Berlin.

"On interviewing Mr. McLean at his hotel, he readily plunged into the whole question relating to health and disease in plant and man with an earnestness characteristic of a genuine, experienced Scotchman.

"Why do you so pointedly couple plants and men together with reference to health and disease?" said the reporter.

"Simply because the one law governs the welfare of both, and man is but a migratory plant, subject to decay, disease and death, from precisely similar conditions, the only apparent difference being in the methods by which plants and men are nourished from the soil and atmosphere. Humanity, in Scripture, is constantly compared with a plant or tree, and we know that He who speaks through the Scriptures is the Creator of both, and knows far better than human wisdom can comprehend the aptitude of the comparison. More than this, it cannot be too often said that the tree has been placed before us in all it stages of construction throughout countless centuries for no other purpose than to serve as a model and symbol of the human tree. At the time when the embryo tree, or germ, first becomes visible with the aid of a powerful microscope, it appears as a little bag surrounded by the far larger bulk of *starch* or *gluten*, which is to form its food in its earlier stages, and which we call the seed. The seed is in fact a bag of food which is to form the sustenance of the baby tree in its center—that embryo tree appearing as yet merely in the form of a soft egg-like bag or vegetable cell. This cell is the progenitor of the tree, which is composed of innumerable cells of similar appearance built upon and joined to one another like stones in the wall of a house. The first cell, the "Adam and Eve in one flesh," of the vegetable world forms within itself two other cells, dividing itself into chambers by a partition, and each of these chambers has the same power of reproduction as the first parent, out of whose substance they are made. Then this first generation of two beings begin each to subdivide into a second generation of three or four baby cells, which form partitions and stand upon one another in close bond as a second course of living stones. These each bring forth their vegetable children, which take their appointed places, *each by subdivision*, in the living structure. The third generation, the third course in the building of the vegetable house, is filling up the little trunk and root of the young tender tree, as yet almost shapeless to the human eye in its simplicity. But successive generations of these progressive subdivisions and their growth into their appointed size and station gradually reveal the shape and character of that race of vegetable cells which is the progeny of the first parent, and which forms the growing tree.

"The blind man in the parable whom Jesus is represented to have restored to spiritual sight could see 'men as trees walking.' The anatomist well knows that the circulatory system of veins and arteries is like a tree, whose sap is blood, branching from the heart to the lungs and skin, and rooted in the stomach and intestines. The skeleton is like a tree, of which the spinal cord is the trunk, the ribs and arms and head being branches, and the legs the roots. The human muscles are as fibers of the tree, which, as in the plant, brace it, and bend it to the required directions. The windpipe is the trunk of the tree, whose sap and nourishment are air, whose root is in the nostrils, and whose branches are the bronchial tubes, terminating in that network of ever-moving twigs and leaves called lungs, and bringing the air of heaven to nourish and purify the leaves and twigs of the venous tree, which is interlaced with it. The skin is the mutual leaf surface of another pair of conjoined trees, arterial and venous, each based in different portions of the heart; and the leaf surface of the outer skin exposes the sap of these two trees to the necessary action of the atmosphere. The nervous system is like an inverted tree whose root is in the brain, whose trunk is in the spinal cord, and which ramifies into every portion of the human body, terminating outwardly in myriads of fine nerve twigs in the outspread foliage of the skin. It will thus be seen that the outward skin represents the aborescence and foliage of the osseous, arterial, venous, and nervous trees which make up the greater portion of the human frame—in fact, man is a compound tree, exposing foliage at every part to the action of the air and light and warmeth. And the lungs also are like foliage of a more delicate kind, to expose the interior ramifications of the arterial and venous trees to the action of the ever-flowing, ever-ebbing air. In the leaves of plants are innumerable little pores or mouths which perspire the surplus moisture of the sap or plant blood, which exhale oxygen and inhale carbonic acid gas from the air. In the foliage of the human skin are similar pores or mouths which perspire the surplus moisture of the human blood and respire the gases of the air in much the same manner as the leaves of plants. And as the leaves of plants fulfill their term of office and then die and fall upon the earth, so fall the leaves or scales of the human skin foliage when their work is done. And as the next crop of leaves comes upon the plant to take up the busy work in the ensuing season, so comes the young growth of skin leaves upon the human tree to fulfill their necessary functions.

"Who can look upon these things," continued Mr. McLean, "although not one-hundredth part of the parallel between men and trees has yet been drawn, and fail to to understand what trees composed the Garden of Eden, among which God walked, and what trees are referred to in Ezekiel, where it is said that **no tree in the garden of God was like** unto him in his beauty, so that all the trees of Eden that were in the garden of God envied him? And **how is it possible to imagine** that **the tree of evil fruit,** which Christ said must be cut down, and the sycamine tree, which was **to be planted** in the sea, **and the** fig **tree,** which withered away, were other than human beings?

" Trees feed upon the crude earth through **their** buried roots, and mankind from refined earth in various forms suitably prepared through nature's wonderful laboratory, **and hence** their equal liability to disease and death from certain impoverishing causes, atmospherically and otherwise; for, as nature abhors a vacuum, she also detests corruption, and in order to hasten the transformation of decaying matter into living tissue, we carry myriads of dormant microbes within us, patiently waiting the necessary conditions to commence their scavenging operations. In like manner are all our fruit plants infested with affinitized insect germs, and were **we** not stone blind **we** would long ago have realized the fact also that our suicidal greed for personal gain has brought about the existing terrible results from which vignerons, fruit growers and agriculturalists generally suffer. Impoverished soil and atmosphere all over the earth have furnished the necessary animating conditions for the parasitical hosts who are now in numerous forms busily at work in orchard, vineyard, vegetable, grain, and grazing plots, as well as within very large numbers of human plants, and which can only be successfully combated through the agency of an effective disease-germ-destroying specific and improved environments."

FROM THE HON. J. STERLING MORTON, U. S. MINISTER OF AGRICULTURE

Department of Agriculture, Office of the Secretary.
WASHINGTON, D. C., April 25th, 1893.

JAMES McLEAN, **M. D.,**

Dear Sir—Your letter of the 15th instant, with accompanying inclosures, is received and I have given to them such consideration as the pressure of administrative duties would allow.

The importance of the subject matter of your letter—the relation of forests to health, to metereological conditions, and to the industrial interests of the world, cannot be over-estimated. We are but beginning to understand and appreciate the influence of the forests not only upon physical conditions, but upon the whole round of human life, including its esthetic and moral aspects. Therefore I hail with hearty welcome every one who has any understanding of the subject and is moved to use his knowledge for the public good. If your proposed remedy for the ravages of noxious insects shall prove practically effective you will have your reward in the grateful thanks of multitudes.

Respectfully yours,
J. STERLING MORTON,
Secretary.

Justification for an appeal to every nation on the momentous question of Forestry:

" Civilization has progressed to the point that makes the great nations of the world amenable to reason."—SENATOR STANFORD. (*S. F. Examiner, May* 3, '93).

SPECIFICATION AND PLAN
Of Insect Pest Eradication Discovery.

The object of my invention is to provide simple, cheap and effective means for the total and permanent extirpation of all insect and parasitical pests injurious or fatal to vegetable growth wherever such have made their appearance, and the prevention thereof in localities where they are still unknown. Especially have my efforts been directed in devising the same, to finding a specific for the extermination of the *phylloxera vastatrix*. (The grape-vine phylloxera or vine pest) and a practical way of preventing its reappearance once it has been rooted out.

I attain this object first by creating a parasitic repellant atmosphere through the agency of Eucalyptus fringes and belts bounding and intersecting generally all cultivation paddocks, as also orchards, vineyards, olive-yards, hop fields, flower and vegetable gardens, and nursery grounds. But while things are being shaped to bring about atmospherical conditions that will completely and permanently bar out all forms of insects inimical to cultivated parts, I have to and do provide a treatment for vegitable productions that have already been attacked, which will stamp out any disease they may be subjected to and relieve their weakness. This treatment consists in the scientific application of the invigorating force and curative properties of electricity, coupled in some instances with the use of a certain chemical compound adapted to rid the growing wood or vegetable matter of its parasitic enemies and favour the benign influence of the electrical power.

That gardens, vineyards, orchards, grain fields, etc., can be thoroughly protected from the disastrous inroads of grass-hoppers, locusts, the codlinmoth, the phylloxera, and every other form of insect pest by the agency of fringes and intersecting belts of *Eucalyptus globulus* and other more hardy sorts of eucalypti properly laid out, and can be cleansed and fertilized by the judicious use of electricity, with or without the aid of an emulsion,—such as I will herein describe—has been demonstrated to my entire satisfaction after years of patient research and carefully conducted experiments. There is produced as a result of the pungent emanations arising from eucalypti surroundings a strong antiseptic atmosphere within which no insect plague can or will exist whilst honey making bees thrive amazingly, yielding an abundant supply of rich medicated honey. Such fringes and belts are also believed to be the best possible substitute that can be provided for the natural barriers formerly afforded by forests, the indiscriminate destruction of which, in connection with impoverished soil, is, in my opinion, the real cause of destructive insect invasions and growth. They also form beneficent shelter from the spread of hoar-frost and severe wind storms, in addition to providing most desirable quarters for insectiverous birds, and affording several other advantages of an economical and health promoting nature. As to the effects producible by electricity in its various states and rates of motion, I have come to the conclusion that an electrical current conveyed by such simple means as are hereafter described may be advantageously utilized as a fertilizer and insect destroyer. By acting on the soil and the water thereby raising their temperature a little higher than that of the surrounding air will change their abnormal condition and enrich the earth's productiveness, impart new life and vigor to the growing plant and annihilate every form of parasite thereon. No one need hesitate in drawing upon the illimitable supply of this real vitilizing and purify-

EUCALYPTUS VILLA. Australia.

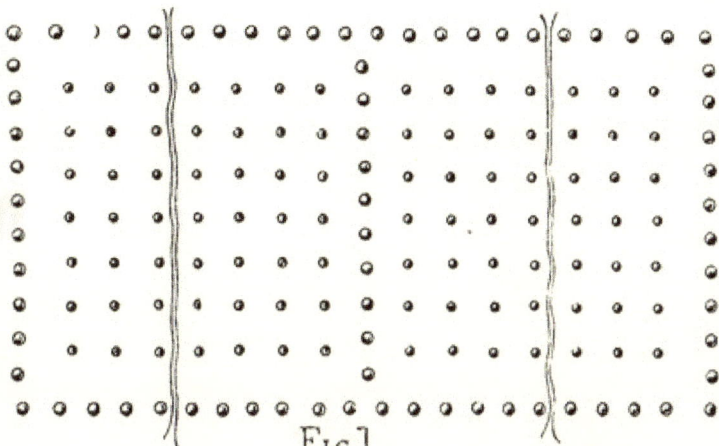

Fig.1.

Fig.2.

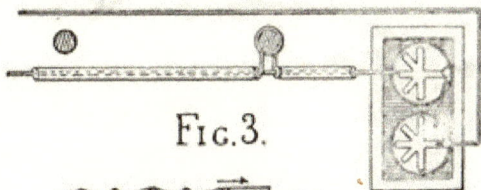

Fig.3.

Fig. 4.

Fig. 5.

ing force, as practical tests will always show the wonderful fertilizing power of electricity through impoverished soil, and its fatal effects on parasitical pests—" nature's scavengers"—including the phylloxera " (in as effective a manner as Professors Mengarini, Bernardi, Martinotti, De Meritens, Rivière and Tolomei's electric methods to destroy undesirable fermenting germs in wine. The latter professor experimented, during 1891, on the action of an induction current on the fermentation of the must, and came to the following conclusion:—

1. The development of *Saccharomyces ellipsoideus* (peculiar **insect pests**) is largely prevented by the action of the electric current, and when the latter is strong enough to produce light in a dark room its development is stopped.

2. The liquid that has undergone the action of the electric current keeps well without developing any fermentation, just as if it had been boiled for some time.

3. The ferment (i. e. the bacteria) is destroyed by a strong electric current." —Professor E. W. Hilgard, California University.)

" Batteries for instance, may be safely and effectively used to cleanse and vitilize all vineyards irrespective of extent or position. Similar treatment is applicable to orchards, and, with some modifications to cultivated lands in general. But the electrical action has to be helped sometimes by the employment of an emulsion—aforesaid. What this emulsion consists of and how it is applied, when and how to treat the diseased, weak or poorly-fed vegetable productions electrically, and in which way to set out and grow the fringes and belts of eucalypti, I shall now proceed to describe in detail,—reference being had to the accompanying drawing and to the letters and figures thereon, which form part of this specification.

Figure 1 of said drawing is a broken plan of a vineyard surrounded and divided by fringes and belts of eucalypti, showing what I consider a suitable way of carrying out my invention.

Figure 2 is a broken sectional elevation giving a practical illustration of the manner of conveying an electric current to a row of vines or other plants.

Fig 3 is a top view showing how the wires from battery may be connected with and disconnected from the vines or other plants undergoing treatment.

Figure 4 is a detailed view of insultating tubes which may be used where the electric wires rest against vines or other plants that no longer need to be electrified.

Figure 5 is a diagramatic view of the electric supply fluid conveying wires, and plants in the circuit, showing how several rows of vines or other plants may be treated with electricity from one and the same source.

A represents vines, which by preference, are set out ten feet distant from one another, in parallel rows about the same distance apart. B and C respectively designate rows of eucalyptus trees disposed in fringes and belts round and through the vineyard, the trees in each fringe and belt ranging from three to eight feet and upwards apart, as determined by circumstances. Each row of eucalypti is about twelve feet distant everywhere from the nearest row of vines, and a distance of about one hundred and sixty feet intervenes between each belt, as also between the end fringes and the nearest belts. The number of intersecting belts C is, of couse, determined by the size of the vineyard, which, as the lines C. 'C. 2 indicate, can be of any length. In laying out the fringes and belts, the plants —Eucalypti rootings—are set in properly prepared trenches at least four feet deep and if possible filled in round plants with a mixture of sand and earth. When planting, the roots should be given a longitudinal and down-

ward set so that they may not run inwardly to interfere with the growth
of the vines or other produce being raised in the plots surrounded by the
eucalypti. During the first year or two the young eucalypti plants should
be protected from extreme heat and cold by the agency of light rough
sapling frames over which any cheap, coarse calico fabric may be secured.
They should also be occasionally examined and certain caterpillars, which
frequently lodge under carefully folded leaves, picked off and destroyed.
Liquid manure applied round their butts will lead to a vigorous growth.
As the plants mature all necessity for special attention decreases, but the
shading bark should be carefully removed and the trees lopped at an altitude of sixty or eighty feet. Then the work of protection is completed
And it may be added that the vineyardist who will carefully follow the
foregoing instructions and raise fringes and belts of eucalypti trees around
and through his property, as directed, need not long fear the ravages of
phylloxera or other voracious pests, as within a short time his estate will
become for quite thirty feet round and beyond its boundaries absolutely
exempt from such insects as may have previously infested it—including
locust invasions, and the more reckless who will occasionally venture within the invisible boundary line extending over the tree tops will drop dead
before reaching the coveted plants.

It will be understood that the spaces intervening between the fringes
and belts of eucalypti and the vines may be utilized for raising beet roots,
tomatoes, and other vegetables, which will grow luxuriantly on at least
eight feet of said space, so there is no occasion to fear any waste of ground
resulting from the adoption of the plan and mode of culture now proposed.
It will be seen also that what has been said regarding the protection of
vineyards, applies with equal force to orchards and other cultivated lands,
with only such changes as difference of culture will naturally suggest.

Having shown how cultivated grounds may be rendered free from insect
pests and afterwards permanently kept exempt from such by growing
round and across them eucalypti fringes and belts that will act like so
many walls and partitions to repel and prevent the attacks of all parasitical pests, there remains to be explained how infected plants may be rid of
the disease affecting them, how others may be preserved in a healthy condition until such time as the protecting fringes and belts shall have come
to maturity, and how the weak ones may at any time be reanimated and
strengthened. As already pointed out, the purifying, preservatory, vitalizing and invigorating agent relied on to do this is electricity, with or without the aid of a chemical emulsion.

As to the apparatus required for conveying an electric current to the plants
and charging them with electricity, none, it is thought, will answer the
purpose better than the simple appliance illustrated in the accompanying
drawing and which merely consists of one or more batteries D with cop-
per wires $E\ E\ 1$ attached, respectively, to their positive and negative poles.
By preference, the wire E leading from the battery or batteries is insulated
by means of an india rubber tube F and passed through the crotch of the
vines or other plants to be electrified—or suspended therefrom, at about
one foot from the ground. Thence it descends through the ground, E
about the top of the roots where it is attached to the bare return wire E, 1
which is connected by loops or short lengths, as at A, with the trunks of
the various plants. The upper wire may be arranged and connected with
the plants in the same manner as the lower one, if desired. Both wires
may also be connected by separate wires running along the stem of any
vine or other plant, to intensify the action of the electric current, when-

ever it is deemed advisible. Infected and other vines as also fruit trees may thus be vitalized, and insect pests—including the phylloxera destroyed by a current of electricity derived from any suitable battery, dynamo, or storage supply. If all the vines in a row are to follow the same treatment, the electric wires may be run and attached thereto in the manner illustrated in Figure 2 of the drawing. If some of the vines are in a healthy state, they may be left out of the circuit and the current diverted to the weaker or more sickly ones, as represented at Figure 3. In such case, the connecting loop or length of wire is dispensed with or removed and the main wire passing by the healthy vines may be completely insulated therefrom by means of an extra over-lapping tube or rubber sleeve *F* covering the part where the connecting loop or length is usually attached. Such an arrangement is shown at Figure 4. Figure 5 shows another arrangement whereby several tiers or rows of vines may be charged with electricity from a single battery or series of batteries. In the latter figure all the electric wires start from the positive pole of the battery and return after winding round the vines in the circuit to a root *G* connected with the negative pole. This is thought to be a most economical as well as effective way to apply the electric current.

While infected vines and fruit trees are purified, vitalized and strengthened by the application of electricity, they will be found to be still more benefited and more completely cured if treated at the same time with an emulsion composed of potash, flower of sulphur and water in the following proportions—potash 2 lbs.; flower of sulpus 3 lbs.; water 5 gallons. The potash is disolved with a little linseed oil; the sulphur and water are boiled twenty minutes in a covered boiler; then all the ingredients are mixed and stirred together. The boiler must be kept well covered and wood only used to stir the compound.

When this emulsion is used in the treatment of vines, the earth around the butts should be well puddled with it to beneath the upper roots and all infected leaves sprayed therewith.

To cleanse blight diseased orchards, the following directions should be followed:—

1. Carefully clear away the earth from round the buts of infected trees for a distance of about one foot from the ground surface down to the top of the upper roots, and with a soft brush liberally paint round the buts from the roots to about fifteen inches above the ground with the emulsion, also moistening the top roots and round the bottom of the opening with the mixture.

2. Leave the cleared out parts open, and with a suitable hose or hand watering-can saturate the ground with water round the opening for a distance of about eighteen inches, avoiding the opening.

3. Should the upper parts of the trees be blighted or otherwise affected, spray such with the emulsion. One application will suffice, but may be repeated, as the emulsion has a specially nourishing effect on vegetation.

4. On day following examine the moistened parts round the openings for dead or dying insects. Repaint the buts and root tops. Also moisten the bottom of openings, then fill in openings with roughly powdered charcoal, over which place a liberal supply of eucalyptus leaves held in position by rough little sapling triangle frames or by short cuttings of saplings.

During a locust invasion, vines, olives and fruit trees may be thoroughly protected until the said fringes and belts of eucalypti mature, by spraying them with the above described emulsion, aided by the eucalypti leaves round the buts.

Having described my invention, what I claim as new, and desire to secure by letters patent is:—

1. The herein described method of driving and keeping destructive insects away from cultivated ground, which consists in surrounding and dividing said ground with or by means of fringes and intersecting belts of eucalypti, substantially as set forth.

2. The herein described method of freeing cultivated ground from insect pests, which consists in planting eucalypti round and through said ground, and charging the vegetable productions therein with electricity, substantially as set forth.

3. The herein described method of ridding cultivated ground of insect plagues, which consists in providing fringes and intersecting belts of eucalypti around and across said ground, and treating the vegetable productions therein electrically with the aid of an emulsion, substantially as set forth.

4. The herein described method of planting fringes and belts of eucalypti round and through cultivated ground, giving the roots of each plant a longitudinal and downward set, substantially as set forth.

5. The herein described method of cleansing, vitalizing and strengthening vegetable productions, which consists in the application of electricity to growing wood or vegetable matter, substantially as set forth.

6. The herein described method of treating plants electrically, which consists in the use of any suitable means of conveying a current therefrom along, around, or through said plants, substantially as set forth.

7. The herein described method of purifying and fertilizing cultivated ground and the vegetable products therein, which consists in surrounding and intersecting the same with eucalypti fringes and belts, substantially as described, and conveying an electric current thereto by means of wires connected with any suitable electric source, aided by an emulsion composed of potash, flower of sulphur and water prepared and applied substantially in the manner and proportions herein set forth.

IMPORTANT.

The various detailed methods in the above specification are so closely interwoven with each other for the eradication *and prevention* of destructive insect plagues as to necessitate their united adoption, in order to ensure the desired results. The electric appliance being to cleanse insect infested roots and to vitalize the surrounding soil, whilst the Eucalypti fringes, etc. repel periodical visitations of grasshoppers, locusts and other destructive insects, as also fungi creating malaria, besides providing exceptional evergreen shelter from wind storms, hoar frosts and pleasurable protection to insectiverous birds. The emulsion, etc., to be used when required until the Eucalypti trees mature. The particulars regarding the trenching and arranging of Eucalypti plant-roots to be simply considered as a detail to ensure said roots from straying in quest of more congenial soil. I therefore—for the above reasons and those stated throughout this treatise —respectfully submit that the plan as a whole is worthy of serious consideration, as by its general adoption a more than gradual restoration of local forest-lungs and consequent climatic benefits within the boundaries of all fruit and grain growing centres would be assured, whilst the harnessing of reinvigorated water-falls and of the ocean beach surf—now seriously contemplated—would furnish an abundance of the electric motive and fertilizing fluid at a nominal cost to rural districts.

ADDENDUM.

FOREST ANNIHILATION IN CALIFORNIA, A STUDY.

(From the S. F. Morning Call, May 28th, 1893.)

"The forests of California are among the wonders of the world, and the lumber industries they give rise to are among the most important on the coast. The chief growth is the redwood, which yields so enormously that the State Board of Forestry estimates that redwood forests comprise half the timber in California, and this though the redwood is confined to the coast, while the area of timber lands in the interior is many times more than that on the coast. Other trees used in the lumber trade are sugar pines, often found 8 feet in diameter and of immense height without flaw; willow, cottonwood, sycamore, oaks of all kinds (including chestnut oak, whose bark tans the famed California sole leather), laurel or bay wood. Redwood burl is an excrescence growing on the sides of the tree, which makes elegant veneering.

"In the United States there are 466,000,000 acres of timber land exclusive of Alaska. Of these 53,000,000 are divided among the Pacific States. These do not include unmerchantable timber. California has a forest area of more than 18,-000,000 acres. Some idea of the volume of the lumber industry may be gathered from a brief presentation of figures regarding it taken from a single representative county—Humboldt. Of 938,000 acres in Humboldt County classified as timber lands about 538,000 were originally covered with redwood forest, leaving about 400,000 acres of other timber divided about equally among pine, spruce, fir and cedar lands, and lands covered principally with madrone and laurel and tanbark, and white, black and live oaks. Of the redwood lands some 39,500 acres have now been cut and sawed into about 4,000,000,000 feet of timber, leaving 498,500 acres standing, which at the conservative estimate of 100,000 feet per acre will produce 49,850,000,000 feet of lumber. The present rate of cutting is about 200,000,000 feet of lumber per year.

"From the port of Eureka the lumber fleet, together with the passenger steamers, took away during 1891, 152,517, 613 feet of lumber, which includes shakes, shingles, pickets, etc., valued at $2,897,834. Of this amount 9,998,663 went to foreign ports, as follows :

To Honolulu	3,937,193
To Sydney	3,796,644
To Guaymas, Mexico	450,161
To La Paz, Lower California	259,852
To Valparaiso	332,336
To Callao	284,007
To Victoria, B. C	182,679
To Central America	246,840
To Tahiti	208,951

"Fresno is another large lumber center, as is Mendocino, and altogether it has been estimated that the annual production of lumber in the State is not less than 300,000,000 feet. Dr. Kellogg, in his " Forest Trees of California," says that " probably from a fair estimate of the redwood along our coast it would not comprise more than 3000 square miles of forest land."

"The amount of timber now standing has been variously estimated, rating all the way from 25,000,000,000 to 100,000,000,000 feet board measure. While in some sections the land will not yield more than from 10,000 to 15,000 feet per acre, there are others which will yield from 250,000 to 500,000 feet, so it will be seen how difficult it is to figure the total closely. As previously indicated, the redwood belt is located on the coast of the Pacific Ocean, and between it and the interior of the State lies the Coast Range. For this reason the railroad touches it at only one or two points, and almost the entire product is transported by water. Both steam and sailing vessels are used for this purpose, and the capital employed in the lumber-carrying trade is a very important factor in the commercial interests of our State.

" There are about forty mills engaged in cutting redwood, the largest having a capacity of 75,000 to 80,000 per day. Perhaps the average capacity of them all would be about 50,000 feet per day. (To multiply 50,000 by 40 totals 2,000,000 feet per day, and by 300 working days in the year, make six hundred millions of feet per annum, for forty years at least forming a grandtotal of 24,000,000,000 feet of redwood alone, exclusive of pine, spruce, fir, oak and cedar removed by lumbermen and forest fires, etc., minus any replenishing effort!) There was manufactured and shipped from the redwood mills in Mendocino and Humboldt counties during the year 1891 about 230,000,000 feet. Of this about 12,000,000 went to foreign countries, while the balance, 218,000,000 was consumed in the Pacific States or shipped to interior States.

"The State Board of Forestry says: "The water flowing in California rivers is more precious than the gold lying hidden in their sands. So long as the forests cover the mountain sides the streams will flow with some evenness throughout the year; but when the forests disappear the rivers will become rushing torrents in the spring and dry arroyos all the rest of the year. The forests of the Sierra Nevada are the natural reservoirs for irrigation of the San Joaquin valley. Hitherto the mountains have been left to the sheep-herder and the millman, who have wrought destruction unheeded and unchecked. Sheep-raising and timber-cutting are legitimate pursuits and entitled to fair treatment, but as conducted in California for many years they have not been conducive to the general welfare. The millman has slashed the forests recklessly, wasting more than he used and not confining his operations to his own property. The sheep-herder, caring only for pasturage, has set fire to the brush annually, burning off the young growth and killing the large trees. The seedlings and shoots that escaped the forest fires were destroyed by the sheep. And so not only has the mature forest been greatly injured but the total extinction of the forest growth made inevitable unless the work of devastation be stopped."

A TARDY YET OPPERTUNE PROTEST.

(*San Francisco Examiner May 28th, 1893.*)

[Special to the EXAMINER.]

WASHINGTON, May 27.—Commissioner Lamoreux of the General Land Office to-day rendered one of the most important decisions that has come from the Land Office in many years, when he decided the famous Redwood land case in the Humboldt district, California. By his decision over 148,000 acres of valuable timber is decided to be the property of the Government on account of the fraudulent entries made by persons who were trying to get this timber from the Government. The probabilities are that the parties who have been defeated in this case will immediately appeal it to the Secretary of the Interior, but now that there is a Democrat in that office he will probably not allow the country to be robbed by speculations, and the decision of the Commissioner will doubtless be affirmed. It is also probable that in the near future this entire tract, with, perhaps, additional lands adjoining, will be set apart as a forest reservation to prevent its dispoilation to reserve it for future use. Assistant Commissioner Bowers of the General Land Office has been all through that country and he recognizes the importance of forest preservation, and when the Commissioner's decision was brought to his attention to-day he at once saw the importance of having the President issue a proclamation reserving this land from further encroachments and depredations. The redwood land case has created a great deal of interest in the West, and the decision of the Commissioner will no doubt be gratifying to all persons who sought the preservation of these lands."

In the May number of this year's *Plant Life*, a San Francisco monthly journal devoted to the advancement of "horticulture, viticulture and floral vegetation," appeared an interesting article under the heading of "Need of a Higher Education," from which I quote the following :

"Plant diseases are not only alarmingly increasing in general, but the affections are becoming more serious in their characters. With these facts prominently presenting themselves, it should be inferred that there must be a radical deficiency in the knowledge we possess governing the hygiene of vegetation, or how, otherwise, can the increase and deadliness in diseases be accounted for? There is no disguising the fact that the disease of the dropping of fruit from trees, even after a well-advanced stage of growth, is becoming alarmingly prevalent and increasing the area of country affected. The diseases of blight and scabs are invading regions where it was thought they could not exist, and evidently these parasites adapt tnemselves to climate and vegetation readily, while countless tribes of insects are fastening upon every variety of plant life. A general review of the horticultural field will show that notwithstanding the care which has been bestowed by government and state, to prevent the spread of the pestilential characters, the result has been the development of new and more dangerous enemies, without a corresponding diminution in the number of the former."

The real cause should now be apparent to every observing forester.

ERRATA.

The artist inadvertently omitted showing the indicating letters on the annexed reduced plan, which will however be sufficiently understood by reference to the respective figures thereon. The large outer and centre rings of Fig. 1 indicate eucalypti plants, and the double transverse lines denote an extension of vineyard or orchard to any required limit within a suitable distance from boundaries and from each intersecting belt. The arrows of Fig. 2 indicate how the electric current travels from the positive to the negative poles of battery. See plan on page 52 of original experimental grounds and specification on pages 55 to 58.

The word "raised" on top line of page 14 should read "razed."

The words "wher uprooted" in 24th line from bottom of page 15 should read "were uprooted."

On the third last line of paragraph under the heading "Fearful Shipwrecks," should read "odd craft."

The concluding portion of fifth line on page 28 should read "two very great factors."

Certain words on page 29, last line, should read "his hair, beard and necktie."

The word at end of the 18th line from top, on page 43, should read "appeared."

TABLE OF CONTENTS.

	PAGE.
PREFACE	2
The Evolution of Obscure Truths	3
Minute Fungacious Organisms	4
The Chemistry of Creation	4
Forest Influence in Gallilee	5
The Order of Creation	5
"That which hath been is now"	8
Boussingault and Humboldt	7
Reboisement in France	9
A Few of the Fruits from Wholesale Forest Destruction	9
Alarming Dispatches (*S. F. Examiner*)	11
Special Dispatches (*S. F. Chronicle*)	12
A Terrible Storm	15
Cursed by Cholera	15
Fearful Shipwrecks	16
Fierce Forest Fires	16
Forest Destroying Combines	17
Dante's Inferno	17
Forest Lands Preferred for Settlement	18
An Australian Conference re Locust Plagues	19
A False Report	20
U. S. Consul E. L. Baker's Report Concerning Eucalypti	21
Further Testimony	23
A Big Bank Failure from Atmospheric Troubles	24
Additional Evidence of Ruined Deforested Soil	24
A Deforestation Lesson from Russia	25
Death Valley	26
Immense Value of the American Grape Growing Industry	28
Other Fruits	28
Locust Plagues in Algeria	29
Killed by Locusts	29
Scripture Warnings	30
Newman's Callista	30
The Vine and Phylloxera in California (Prof. Husmann)	31
French Report on the Bi-sulphate of Carbon Treatment for Phylloxera	33
The Phylloxera in Australia	34
Professor F. W. Morse	35
Concerning Known (Phylloxera) Remedies	36
The Phylloxera Question	37
Comparative Results	37
Vine Troubles in France	38
Atmospheric Germs	39
The Mediterranean Flour-Moth	42
A Wail from Malta	42
Some Capitalistic and Hygienic Consequences of Atmospheric Troubles	44
Poor Crop Outlook	45
California's Deforesting Contribution	45
Dangerous Experiments	46
Conclusion	47
Ruinous Contrasts	48
Remedy	49
Personal	49
The Hon. Sterling Morton and Senator Stanford	51
Plans and Specification of Insect-Plague Eradicating Discovery	52–56
Important and Addendum, "Plant Life, Etc	57–59
Eratta	60

www.ingramcontent.com/pod-product-compliance
Lightning Source LLC
Chambersburg PA
CBHW030717110426
42739CB00030B/712